Felix Publishing 2017
www.felixpublishing.com.au
email: info@felixpublishing.com
Print copies available from publisher.

Riches from the Earth
Part of the Series **Adventures in Earth Science**
Other books in the series include:
> Exploration Science (Field Geology and Mapping)
> Changing the Surface (Erosion and Landscapes)
> Rocks - Building the Earth
> Fossils - Life in the Rocks
> A Dangerous Planet (Earth Hazards)
> Through Sea and Sky
> Beyond Planet Earth (Astronomy)

2016 digital book release
ISBN: 978-0-9946432-5-4
2017 Print Edition
ISBN: 978-0-9946432-6-1

Author: Dr P.T.Scott
All illustrations, photographs and videos by the author unless stated. Cover photo: Needles of the mineral Crocoite (Lead Chromate), Zeehan, Tasmania. Cover design after that of AJS Creative, Brisbane.

Registration:
Thorpe-Bowker +61 3 8517 8342
email: bowkerlink@thorpe.com.au

FELIX
PUBLISHING

RICHES

FROM THE

EARTH

Dr. Peter T. Scott

FELIX
PUBLISHING

First released 2017

To my grandchildren who are
yet to find their own adventures.

About the Author

Dr. Peter Scott is an award-winning teacher of Earth Science of over forty years' experience in both Secondary and Tertiary Education. He holds a Bachelors' Degree, two Masters' Degrees and a Doctorate including many years on his own research in locating and correlating coal measures. For his work with national bodies in the mining industry and in administering and teaching Secondary Education in mining and energy, he was awarded the prestigious Queensland Minerals and Energy Academy Award in 2008.

Dr Scott and students with a bar of pure gold, Stallwell, Victoria, Australia

Table of Contents

Chapter 1: Materials of the Earth

1.1 Introduction

The outer parts of our planet which are within our immediate experience have been divided into three main parts. They are the:

Atmosphere which consists of the gases of the air, mostly nitrogen gas (78% volume), oxygen gas (21%), argon gas, (less than 1%), carbon dioxide gas (0.03%), water (varies), and small traces of other gases, dust and pollen. Gases from the air have been extracted by compression and liquefaction and are used for a wide range of medical and industrial uses;

Hydrosphere which consists of the waters of the Earth which covers over 70% of the planet's surface as ocean, ice caps, fresh water and soil moisture. Fresh water is the world's most valuable resource and its supply is a constant necessity for life on this planet. The oceans are also a valuable source of food, mineral resources from solution and a source of freshwater by way of the water cycle; and the

Lithosphere which consists of the rocks, minerals, soils and weathered products which make up the solid material of the outer parts of the Earth's crust. These include:

- **Minerals** are the basic components of the lithosphere and are naturally-occurring, inorganic (i.e. non-living) substances having definite properties and a known chemical composition. They can be formed by **crystallization** from molten material as **magma** or **lava**, e.g. quartz within igneous rocks, or from hot vapours and solutions e.g. sulfur from volcanic vents, or by evaporation from water solutions e.g. halite or rock salt.

Figure 1.1: Sulfur crystallising from a vent in the Kilauea volcano, Hawai'i

Figure 1.2: Salt being harvested from a salt lake in Bolivia where it crystallised from solutions (Photo: Matthew Scott)

Minerals are also the source of many important materials useful to humankind, including:

- **ore** for extraction of metals e.g. haematite for iron

- minerals for building materials e.g. gypsum for plaster, silica (quartz) for glass, calcite (in limestone) for cement, rocks for building

- essential mineral solutions in soils for plant growth e.g. magnesium solution for green chlorophyll in leaves

- minerals as a source of raw materials for the chemical industry e.g. sulfur for making sulfuric acid, phosphate minerals for fertilizers, nitrate minerals for explosives; rock salt for chlorine production

- gemstone minerals for decoration, industry and trade e.g. diamonds for jewelry and garnet for abrasives

- **Rocks** are made up from two or more minerals and have a wide range of formation and uses. In simple terms, rocks can be formed as:

- **Igneous rocks** from the cooling and crystallization from molten rock material such as magma within the Earth's crust or from lava upon its surface. Igneous rocks formed below the Earth's surface are called **intrusive igneous rocks** and usually have large, well-formed interlocking crystals. These rocks include granite, gabbro and diorite. **Extrusive igneous rocks** are formed on the Earth's surface, usually from volcanic eruption. Some examples are basalt, tuff (a volcanic ash), andesite, rhyolite, andesite, obsidian and pumice.

Figure 1.3: The igneous rocks gabbro (left) showing large crystals and vesicular basalt (right) with no crystals but large gas bubbles

- **Sedimentary rocks** from the deposition, compaction and cementation of sediments eroded and transported from existing material. They can form as **clastic sedimentary rocks** by the compaction and cementation of particles called **clasts**, such as in shales (from muds), sandstones (from sands) and conglomerates (from gravel). They may also form as **non-clastic sedimentary rocks** which do not have eroded particles but are made from chemical deposition or from the organic remains of ancient living things. Examples of non-clastic rocks include limestones, evaporites and coal.

Figure 1.4: the sedimentary rocks sandstone (left) with medium-sized sands grains (clasts) and the organically-formed, non-clastic rock coal (right)

- **Metamorphic rocks** from the changes to all other existing rocks by the application of heat and/or pressure. Metamorphic rocks are often crystalline, very hard and may also have a new range of minerals to that of the original or parent rock. They can be formed as **contact metamorphic rocks** by the action of nearby heat sources such as large igneous bodies and examples include quartzite from sandstone, marble from limestone, and hornfels from shales. They can also be formed as **regional metamorphic rocks** which are formed below the surface by strong pressures and heat within the Earth, which flatten the rock and its constituent minerals, or create new minerals which can exist at high pressures such as micas. Examples of these rocks include phyllite and schists derived from shales, and gneiss from granites and shales. When they are exposed to the surface by uplift and erosion, they usually cover a very large area.

Figure 1.5: The metamorphic rocks quartzite (left) formed from sandstone by contact with heat and a schist (right) formed by pressure on shales.

Rocks have been used from before the earliest recorded history as stone tools and fireplaces. The earliest known buildings in Mesopotamia, Egypt and Greece showed that early civilisations soon had great skills in using stones and had an elite class of stonemasons to apply them. Stone such as sandstone, limestone, marble and granite were all favoured materials to build some of the great buildings and monuments of ancient times.

Figure 1.6: A replica stone axe of greenstone, a metamorphic rock. The axe was made recently by a stone-age tribe living in the mountains of Tana Island, Vanuatu and presented to the author

Figure 1.7: The Pyramids at Giza, Egypt are made of huge sandstone blocks which were covered with a highly polished layer of white limestone, some of which still remains at the top of the pyramid at right

Stone was also carved into statues and useful containers such as vases and bowls.

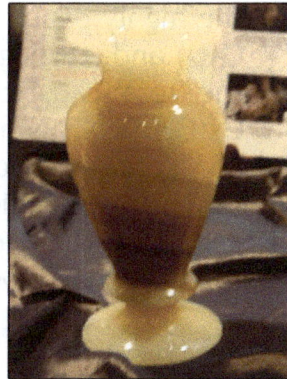

Figures 1.8 & 1.9 At left, Egyptian alabaster, a form of fine limestone, being carved in the old style in a small factory near Luxor, Egypt, and at right, a finished vase of alabaster

- **Clay** is formed from the weathering of feldspar minerals from rocks. It then may be deposited as mudstones or shales which then further weather and breakdown to clay. Muds and clays were perhaps one of the oldest known building materials. Clay bricks, made from shaped mud have been found from before 7500 BC, at Tell Aswad, in the upper Tigris region and in southeast Anatolia, Turkey. Clay was also used from ancient times to make jars and other utensils.

- **Sand and gravel** have long been used in building. Sand, which is mostly silica or silicon dioxide as the mineral quartz, is derived from the erosion of sandstones or from quartz-rich granites. Gravel is derived from eroded conglomerates or from a variety of rocks worn down and rounded in fast streams or on storm beaches. Sand and gravel have long been used as aggregate, the main ingredient in concrete, when they are mixed with lime. Lime is made by strongly heating crushed limestone which is converted into the highly caustic material quicklime (calcium oxide – CaO) and, through subsequent addition of water, into the less caustic slaked lime (calcium hydroxide - $Ca(OH)_2$). This is used as a binding agent in the concrete. The Romans used a special mix which also included volcanic ash which produced a concrete, called by them *opus caementicium*, which was able to harden underwater.

Figure 1.10: The Pantheon in Rome is a good example of Ancient Roman concrete construction

Pure white silica sand is melted to form glass. This is usually done with added limestone and soda ash (sodium carbonate $Na_2(CO_3)$) to reduce the melting point of the silica. Evidence suggests that the first true glass was made in the Middle East and Egypt about five thousand years ago, probably accidently as a beads formed as a by-product of metal **smelting**.

Figure 11: An example of Egyptian glass of the 5th Century A.D. (Photo: WikiCommons)

- **Soils** are formed from existing rocks which have been weathered on site by the reaction with water, gases and natural acids and then worn down by the abrasion of wind, water, ice and gravity to smaller fragments. They may also be chemically changed to new minerals, such as iron oxides, and then often mixed with organic remains and other minerals from solution. **Regolith** is a general name for broken and loose material, such as **scree** caused by landslides. More on rocks can be found in another book in this series: ROCKS – BUILDING THE EARTH.

1.2 Minerals and Their Properties

Minerals can be described, classified and identified in terms of their most obvious characteristics or properties:

- Physical properties which refer to the general appearance, strength and energy interactions of minerals which do not concern the change of the mineral into a new substance. Physical properties are the most useful features in describing or identifying minerals as they often require only simple tests and can be done quickly in the field.

- Chemical properties which refer to the interactions which minerals may have with other substances and energy which results in their change to new substances. The determination of a mineral's chemical properties may require more complex geochemical analysis in a laboratory, but a few simple tests such as testing with acid, can also be done in the field.

The most useful mineral properties are:

1. **Habit** describes the overall appearance and arrangement of the crystals of the mineral specimen. The habit is useful as a diagnostic aid when the mineral has an easily identifiable form. A mineral may have several possible habits. The appearance of the specimen depends upon the conditions under which it formed and its chemistry and internal structure. The most common habits are:

- Irregular – crystal faces in many directions e.g. feldspars

Figure 1.12 Orthoclase feldspar aggregate

- Fibrous – clusters of fine threads e.g. asbestos

Figure 1.13: Fibres of green asbestos

- Radiating - star pattern of needles e.g. gypsum

Figure 1.14: Needle-like clusters of gypsum

- Botryoidal - semi-spherical lumps e.g. haematite

Figure 1.15: Botryoidal haematite

- Mammillary - similar to botryoidal but with flatter, rounded lumps e.g. goethite

Figure 1.16: Mammillary structure of goethite

- Pisolitic - pea-like spheres e.g. bauxite

Figure 1.17: Small pisolites within bauxite

- Foliated – with flat sheets e.g. micas

Figure 1.18: Flat sheets of muscovite mica

- Massive - shapeless with no crystal faces seen e.g. kaolinite. Some of these minerals may be **amorphous**, with no true crystal at all

Figure 1.19: Massive kaolinite

In addition, some minerals often show several types of **twinning** - when several individual crystals grow together along a common flat crystal plane. Some forms of twinning include:

- **Simple twinning** is when two crystals grow together in some symmetrical way e.g. side-by-side or as a mirror image e.g. gypsum

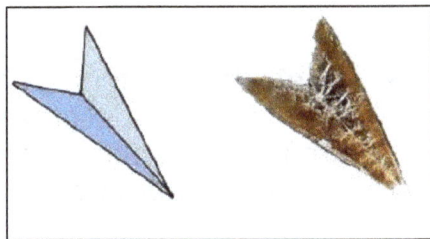

Figure 1.20: Simple twinning in gypsum

- **Baveno twinning** is a form of simple twinning, named after the town in northern Italy. In this form the crystals are joined together in a diagonal pattern e.g. orthoclase feldspars

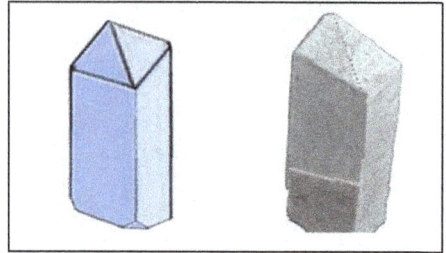

Figure 1.21: Baveno twinning in orthoclase

- **Multiple twinning** occurs when many crystals grow side-by-side many times with a reversal of crystal growth direction e.g. plagioclase feldspar

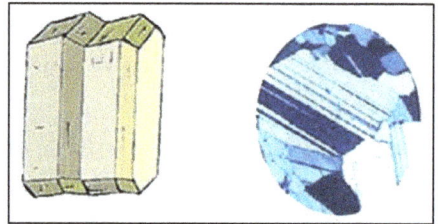

Figure 1.22: Multiple twinning in plagioclase (at right shown in thin-section and seen as parallel lines within the crystal)

- **Penetration twinning** in which one crystal grows through the other at an angle e.g. staurolite

Figure 1.23: Penetration twinned crystal of staurolite

2. **Colour** is shown by the wavelength of light emerging from within the internal structure of the mineral. This may not be reliable in identification, as in many cases a mineral may have different colours under different conditions e.g. quartz could be clear (as rock crystal), white (as vein quartz), smoky grey (as cairngorm), purple (as amethyst), pink (as rose quartz) and other colours depending upon slight impurities of metallic elements.

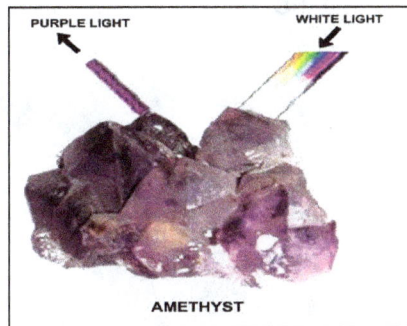

Figure 1.24: Colour as a transmitted wavelength

PURPLE LIGHT   WHITE LIGHT

AMETHYST

3. **Streak** is the colour of the powdered mineral made by firmly pressing the mineral across a white tile called a streak plate. A black tile is used if the mineral gives a white, cream or light yellow streak. Minerals which are harder than the plate have to be scratched by harder minerals, such as corundum or by a tungsten-tipped scribe or file and then the powder smeared across white or black paper. Streak is usually more reliable than colour e.g. haematite may be red, black and various shades of brown but its streak is always cherry-red.

Online Video 1.1: Using a streak plate
Go to https://www.youtube.com/watch?v=Pqt4VsU-JGk

4. Lustre is the way which light reflects off the surfaces of the mineral specimen. Lustre can be:

- Metallic - with a hard, very reflective shine like metals e.g. pyrite (yellow at right), also gold, galena

- Sub-metallic with the reflection similar to metals but not as shiny e.g. sphalerite (black at right)

Figure 1.25: Ore specimen with pyrite and sphalerite

- Non-metallic - a great variety of common lustres including:

  Adamantine (or brilliant) - sparkling like diamond
  Vitreous - glassy e.g. quartz
  Sub-vitreous - not as glassy e.g. calcite
  Pearly - like a pearl or shirt button e.g. talc
  Silky - shiny & fibrous like silk e.g. gypsum
  Resinous - dull but with a plastic-like appearance like resin or plastic e.g. white opal
  Dull - very little reflection; dull like dirt e.g. limonite

Other terms or combination of terms may be preferred. Some minerals may have several different lustres from one specimen to another because of different habits and a particular specimen may have different lustres on different cleavage planes e.g. feldspars may have a sub- vitreous lustre on one cleavage plane but a dull lustre on the other cleavage plane.

5. **Diaphaneity** is the way that light passes through the specimen in normal thickness. This may be described as either:

- Transparent - with light passing through undistorted so that images can be seen through it e.g. topaz

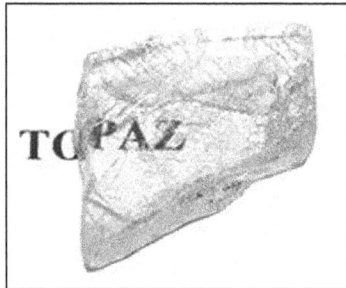

Figure 1.26: Transparent diaphaneity in topaz

- Translucent - light passes through but is distorted like in bathroom glass e.g. muscovite mica

- Opaque - light will not pass through the normal specimen e.g. feldspar.

This property is usually useful only when the mineral is typically transparent or translucent, especially gem minerals such as topaz and chalcedony.

6. Hardness is the resistance to scratching when tested with a standard set of items having a relative strength. The scale developed by Friedrich Mohs (German: 1773-1839) uses a set of standard minerals which have been given values of hardness and arranged in order from softest (1) to hardest (10).

When using this scale, test minerals are firmly pressed across a good, flat surface of the unknown specimen to see if they will scratch the surface. For example, if the mineral fluorite will just scratch the unknown mineral, then that mineral will be said to have a hardness of 4.

When testing hardness, it is a good practice to rub the mark left when the test mineral is pressed across the surface - the mark could be traces of powder left behind by a softer mineral. This will rub off, but a true scratch mark will be left behind. If the hardness of a mineral is between two values, then an half value can be given e.g. if the mineral has a hardness between that of fluorite (4) and apatite (5), then its hardness can be given as 4.5In the field, a geologist

MOH'S SCALE OF HARDNESS		
HARDNESS		MINERAL
1		TALC
2		GYPSUM
3	INCREASING HARDNESS	CALCITE
4		FLUORITE
5		APATITE
6		FELDSPAR
7		QUARTZ
8		TOPAZ
9		CORUNDUM
10		DIAMOND

Figure 1.27: Mohs' scale of hardness

may carry some common items with which to quickly and conveniently test for approximate hardness.

Some items can be used in the field for quickly testing the hardness of specimen. These are shown in Figure 1.28. Whilst some minerals vary in hardness, others, especially the harder gem minerals, have very specific diagnostic hardness. In the field, a geologist looking for gem materials will pick up a piece of white common vein quartz to use to test hardness on any specimen which has a transparent or translucent diaphaneity i.e. it looks like glass.

FIELD SCALE OF HARDNESS	
HARDNESS	INSTRUMENT
2·5	FINGERNAIL
3	COIN
5·5	KNIFE BLADE
6·5	STEEL FILE
7	COMMON QUARTZ

Figure 1.28: A simple hardness scale for field use

Online Video 1.2: Testing the hardness of a mineral
Go to https://www.youtube.com/watch?v=Kvn6u19G8pw

7. **Cleavage** is the way that some minerals split along natural, flat planes of weakness when gently struck. These planes can be:

- perfect - very smooth and shiny

- good - smooth with some shine

- indistinct (or poor) - flat surfaces with some roughness giving little shine

- no cleavage

As well, cleavage may be described as being in:

- One direction (**basal cleavage**) giving small sheets e.g. mica

Figure 1.29: Basal cleavage in muscovite mica

- Two directions giving small steps e.g. feldspars

Figure 1.30: Cleavage in two directions in orthoclase feldspar

- Three directions giving points e.g. galena

Figure 1.31: Cleavage in three directions in galena

- Four directions (**octahedral cleavage**) giving pyramids e.g. fluorite

Figure 1.32: Octahedral cleavage in green fluorite

However, some minerals may grow as well-formed crystals, with faces which look like cleavage planes, but because of their internal crystal structure, these minerals may not have any cleavage at all.

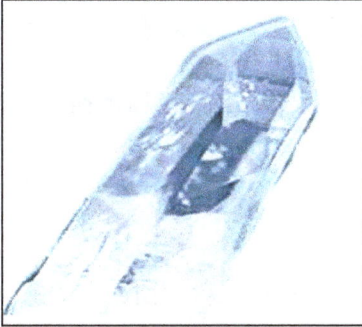

Figure 1.33: A crystal of quartz showing flat crystal faces. This mineral does not give cleavage planes when broken

Determining cleavage by inspection is satisfactory for some of the common minerals which have obvious cleavages, such as those with perfect basal cleavages such as micas and topaz, and the three cleavages of galena, pyrites and calcite. It can become confusing. However, some minerals such as quartz have no cleavage when struck, but do show good flat planes meeting at edges or points due to the original growth of its crystal faces. The only true test for cleavage is to gently strike the specimen and observe the number of non-parallel planes along which it splits.

Online Video 1.3: Finding the cleavage of calcite
Go to https://www.youtube.com/watch?v=_RwRQ3Cgnfg

8. **Fracture** is the way the mineral breaks up roughly, but not along smooth planes of weakness. Not a very helpful diagnostic property in general, but there are some useful patterns of fracture which can be a characteristic feature of a particular mineral. These include:

- even - surfaces are generally flat but not as good as cleavage

- uneven -very rough surfaces

- splintery -long, broken parallel fibres

- earthy - rough but with rounded edges like lumps of dirt

- hackly - rough, pointed surfaces in many directions

- conchoidal – gives very sharp edges with concentric circular marks like the edges of broken glass or shell. **Conchoidal** fracture is the most easily identified fracture seen in the field and is typical of quartz, topaz and volcanic glass.

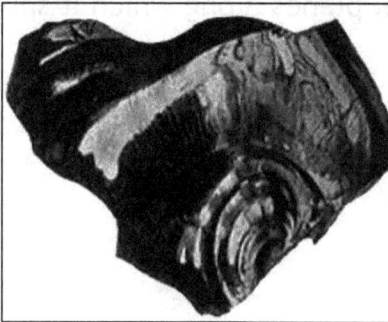

Figure 1.34: Conchoidal fracture in obsidian, a volcanic glass

9. **Crystal Family** is a useful diagnostic aid when some specimens have distinct, well-formed crystals, large enough to show distinct crystal faces and the angles between them.

Usually, the shape of the crystal family is defined by the axes (more than one axis), or the framework lines, around which the outer surfaces or crystal faces are built and the angles between these axes. For example, the simplest crystal shape is a cube, which can be defined as being that shape having all three axes equal in length and being at right-angles to each other (i.e. axis a=axis b=axis c all at 90°).

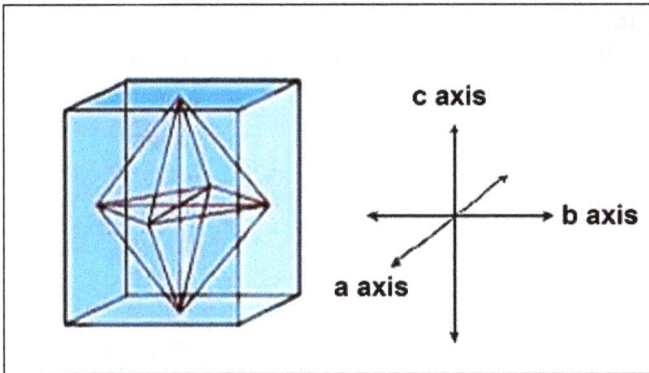

Figure 1.35: The parameters for the cubic crystal family

However, even the simple cube can have complex outer surfaces but still have the same axes.

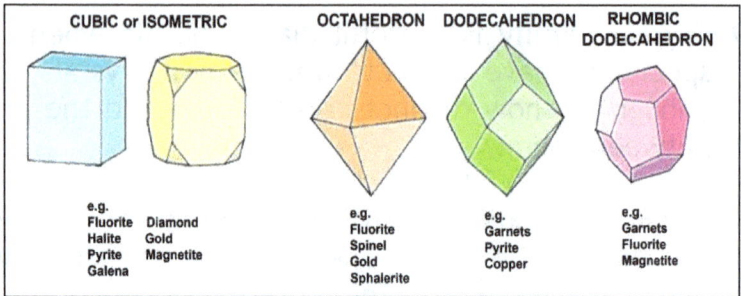

Figure 1.36: Other forms of the cubic crystal family

This identification is only useful if the simple shape of the crystals can be seen and there are no other variations of the crystal form. Complications occur when crystals grow as variations of the simple form or when they are twinned or when they exist as complex aggregates.

Figure 1.37: An aggregate of quartz crystals

In very simplified terms, and for ease of observational identification, the crystal families can be summarised as the:

- **Cubic Family** in which the axes are equal length and are all at 90° to each other e.g.

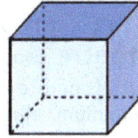

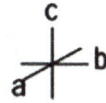

$a = b = c$
all angles = 90°

> halite (salt- sodium chloride)
> galena (lead sulfide)
> pyrite (fool's gold - iron sulfide)
> fluorite (calcium fluoride)
> diamond (the element carbon)
> magnetite (magnetic iron oxide)

- **Tetragonal Family** in which two axes are equal in length and the other shorter or longer, with all axes at 90 degrees e.g.

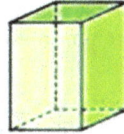

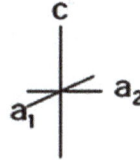

$a1 = a2 \neq c$
all angles = 90°
($\neq$ means "not equal to")

> zircon(zirconium silicate)
> rutile(titanium dioxide)
> cassiterite (tin oxide)
> chalcopyrite (copper iron sulphide)

- **Hexagonal Family** with two horizontal axes equal in length and at 120⁰ to each other and an unequal vertical axis at 90⁰ to these two. This often gives a six-sided crystal in a

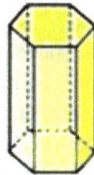

$a1 = a2 = a3 \neq c$
angle between a axes & c = 90°
angles between a axes all = 60°

hexagonal shape, but this may not be the case with some minerals e.g.

beryl (beryllium aluminium silicate)
graphite (carbon)
apatite (calcium fluoro-phosphate)
chile saltpetre (sodium nitrate)
tourmaline(complex silicate with boron, aluminium, alkali metals iron and magnesium)

- **Orthorhombic Family** in which all axes are unequal in length but are all at 90^0 to each other e.g.

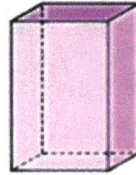

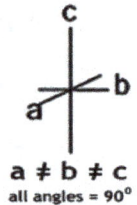

$a \neq b \neq c$
all angles = 90°

aragonite (calcium carbonate)
barite (barium sulfate)
olivine (iron-magnesium silicate)

- **Monoclinic Family** in which all axes are unequal in length with one axis vertical, another at 90^0 to the vertical and the third axis at an oblique angle to the plane of the other two axes.

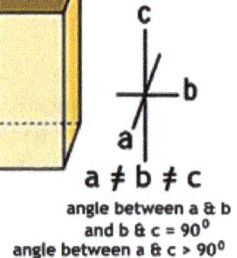

$a \neq b \neq c$
angle between a & b and b & c = 90°
angle between a & c > 90°

muscovite(complex aluminosilicate)
biotite (complex alumino-silicate)
orthoclase (potassium alumino-silicate)
gypsum (calcium sulfate with water)
hornblende (complex aluminosilicate)
augite (complex alumino-silicate)
malachite (basic copper carbonate)

26

- **Triclinic Family** in which all axes are unequal and none of the angles are at 90^0.

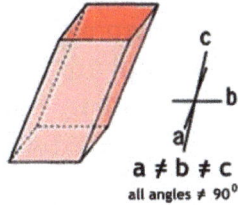

$a \neq b \neq c$
all angles $\neq 90^0$

plagioclase (calcium to sodium alumino-silicate),
kaolinite (aluminium silicate),
rhodonite (manganese silicate).

Some 3-dimensional classifications use **crystal system** rather than crystal families. This is a more technical classification and uses a complex identification, involving planes, or mirrors of symmetry and amounts of rotation around each axis as a rotational symmetry. Consequently there is often considerable confusion when discussing types of crystal shapes. For example, there are six crystal families and seven crystal systems, the latter including an additional member, the trigonal system:

- **Trigonal Crystal System** with all axes equal and all angles equal (but not 90^0). Some members show external hexagonal shape (e.g. quartz) and others a rhombic shape (e.g. calcite). It is a subset of the Hexagonal System.

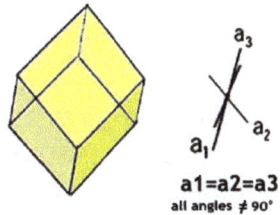

a1=a2=a3
all angles $\neq 90°$

quartz (silicon dioxide)
calcite (calcium carbonate)
dolomite (calcium magnesium carbonate)
corundum (aluminium oxide)

The term **Rhombohedral System** is often used synonymously for the trigonal system, but the rhombohedral system is defined by its crystal lattice rather than outward appearance and is a subset of the trigonal lattice system. It has all sides equal including the c-axis, unlike the hexagonal system where the c-axis is a different length and the angle between the c-axis and the others is not 90^0. Calcite and dolomite can sometimes have this shape. A summary of the main terminology used in crystallography is given below.

CRYSTAL FAMILY	CRYSTAL SYSTEM	LATTICE SYSTEM
Triclinic		Triclinic
Monoclinic		Monoclinic
Orthorhombic		Orthorhombic
Tetragonal		Tetragonal
Hexagonal	Trigonal	Rhombohedral
		Hexagonal
	Hexagonal	
Cubic		Cubic

Table 2.1: Different systems of crystallography

10. Specific Gravity (S.G.) is the density of the mineral compared to that of water which is 1.0 gram per cubic centimetre.

i.e. **Specific Gravity** = $\dfrac{\text{density of mineral}}{\text{density of water}}$

The density of the mineral is calculated by dividing its mass, usually in grams, by its volume in cubic centimetres (note: 1 c.c. approximately equals 1 millilitre). Alternatively, this can be found by dividing the mass of the mineral weighed in air by its apparent loss in mass when completely submerged in water using **Archimedes' Principle**. Since it is a ratio, specific gravity, has no units of measurement e.g. the S.G. of gold is 19.3.

Heft is a rough comparison of the heaviness of a mineral which may be useful for quick identification of some minerals. Heft may be ranked as heavy, medium or light e.g. calcite has medium heft, but barite, which it closely resembles, has a heavy heft.

11. Chemistry - refers to the chemical composition of the mineral i.e. what elements and chemical bonding make up the mineral. This can be found out by:

- chemical analysis in the geochemistry laboratory using standard chemical tests to find the composition of minerals.

- x-ray crystallography in which X-rays are beamed into a crystalline powder and the diffraction or bending of the rays passing through the crystal lattice produces a pattern on photographic film which is typical of that crystal system.

- spectroscopy in which a powdered sample of the mineral is burnt in a gas flame and the colour of the emitted light is observed through a spectroscope which disperses the light into separate colours typical of the chemical elements within the specimen. Computer analysis of the flame colour gives the chemical elements and an approximation of their abundance within the mineral.

Some simple chemical tests can be applied in the field to identify some common chemical families e.g. acid on carbonates such as calcite, dolomite and magnesite, will give bubbles of odourless carbon dioxide gas. Strong acid on sulphides, such as pyrite and galena, will give a bad odour of rotten eggs due to hydrogen sulfide gas and this test can be used to distinguish fool's gold (pyrite) from real gold which has no reaction with the acid.

Figure 1.38: Acid on any carbonate mineral will give bubbles of odourless carbon dioxide gas

12. Specific Properties - Some minerals may have some specific or unique features which may be useful in quick identification. e.g.

- magnetite can be attracted to a magnet and may act as a magnet itself as the variety lodestone

- talc feels slippery

- micas are flexible and elastic

Figure 1.39: Magnetic properties of lodestone

- pitchblende, uraninite and monazite are radioactive

- kaolinite has an earthy taste

- halite (rock salt) tastes salty

Figure 1.40: Green fluorite gives a purple glow in ultraviolet light

- fluorite, autunite and some other minerals are fluorescent in ultra-violet light

- tourmaline becomes charged electrically when heated

- Iceland spar is a clear form of calcite which gives double refraction i.e. splits light up into two beams

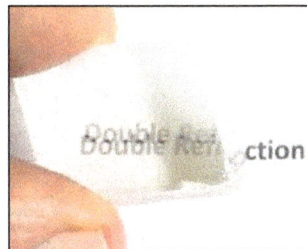
Figure 1.41: Iceland spar showing double images by double refraction

In the field looking for minerals and gems, the geologist usually uses a small range of common properties such as colour, hardness and habit but most minerals in the field are also often badly changed by the weather so the task becomes harder.

Chapter 2:   The Chemistry of Minerals

2.1 Introduction

Geochemistry is a major sub-division of geology. It involves the testing and determination of the chemical composition of minerals, i.e. what elements make up the mineral and how they are arranged together. This often involves complex laboratory equipment and tests.

Most of the common minerals have been formed from only a few of the elements of the Earth. The most common natural chemical elements are given in the chart below:

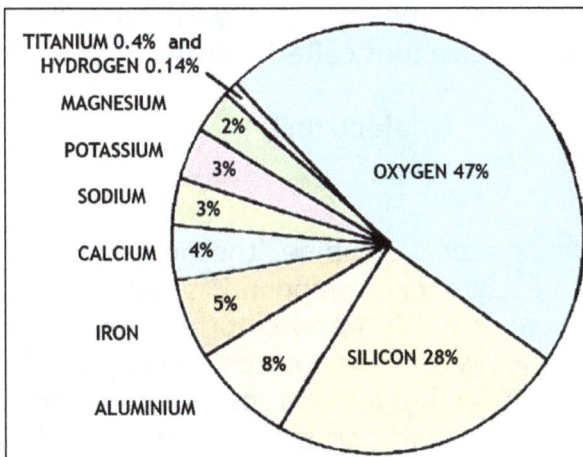

Figure 2.1: Main chemical components of the Earth

Atoms are the simplest forms of the 92 natural chemical elements. In minerals, which are compounds of two or more elements chemically combined, atoms of these elements may do so by sharing electrons, losing or gaining electrons. These are particles in the outer shells of atoms which have a negative electrical charge, and when they are lost or gained, they form charged atoms or groups of atoms called **ions**. For example, the sodium atom has 11 electrons, each with a negative charge, in its outer shells and 11 protons each with a positive charge inside its central nucleus. Metals such as sodium (chemical symbol: Na) tend to lose electrons to form positive ions called **cations**:

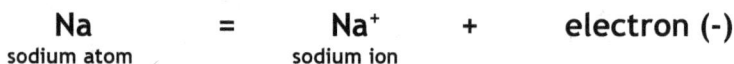

Na	=	**Na$^+$**	+	**electron (-)**
sodium atom		sodium ion		

Similarly, the chlorine atom has 17 electrons and 17 protons and as a non-metal, tends to gain electrons becoming negative ions called **anions**:

Cl	+	**electron (-)**	=	**Cl$^-$**
chlorine atom				chorine ion

When ions or atoms combine, the overall charge on the new, stable chemical compound is zero. For example, when sodium ions (Na$^+$) and chloride ions (Cl$^-$) combine as when seawater evaporates, crystals of sodium chloride (salt = the mineral halite) form so that the charge on the salt crystal is zero, and simplest representation, its empirical formula, of the crystal containing a large number of these ions is:

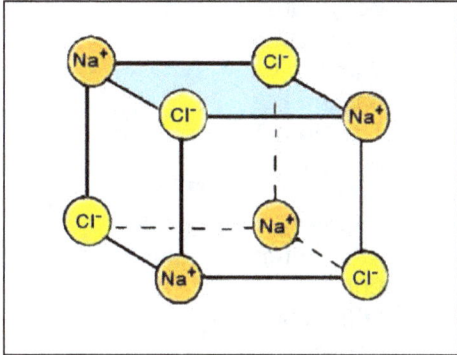

Na^+ Cl^- ionic charges   (+) + (-) = zero charge

Figure 2.2: The simplest representation of a cube of sodium chloride

Note that chemists use symbols when writing the chemical formula of compounds such as minerals. However, a mineral may be a complex mixture of several pure chemical compounds and so have a range of chemical properties e.g. the mineral olivine is a mixture of iron silicate and magnesium silicate in various proportions.

When oppositely charged ions electrostatically come together, a strong bond called an **ionic bond** is formed. When they do this, they form as **crystal lattices**, which are large frameworks of ions held together in the best possible shape depending upon the charges, size, and the directions of the bonds.

Lithium Li^+ Fluorine F^-
Sodium Na^+ Chlorine Cl^-
Potassium K^+ Oxygen O^{2-}
Hydrogen H^+ Sulfur S^{2-} but
Beryllium Be^{2+} variable
Magnesium Mg^{2+}
Calcium Ca^{2+}
Barium Ba^{2+}
Titanium Ti^{2+}
Manganese Mn^{2+}
Iron Fe^{2+} or Fe^{3+}
Copper Cu^{2+} or Cu^+
Aluminium Al^{3+}
Chromium Cr^{3+}
Silicon Si^{4+}
Carbon C^{4+}

Figure 2.3: Some of the most common ions

Sometimes these ionic charges can be moved within the crystal lattice. For example, tourmaline crystals can be given an electrostatic charge upon heating, and quartz crystals can produce a small current when physical pressure is applied. This is the piezo-electric effect used in automatic gas-lighters.

In most minerals, atoms may combine to form complex ions (or groups) with the overall combination of atoms having an excess charge such as:

Silicates $(SiO_4)^{4-}$ e.g. olivine $(Fe^{2+} Mg^{2+}) SiO_4^{4-}$
Carbonates $(CO_3)^{2-}$ e.g. calcite $Ca^{2+}CO_3^{2-}$
Phosphates $(PO_4)^{3-}$ e.g. apatite $(Ca^{2+})_5 (PO_4^{3-})_3 (F,Cl,OH)$
Sulfates $(SO_4)^{2-}$ e.g. gypsum $Ca^{2+}SO_4^{2-}. 2H_2O$
Hydroxides $(OH-)$ e.g. gibbsite $Al^{3+} (OH^-)_3$

Note that the charges all add to zero in each formula in the list above e.g. in apatite there are 5 x 2+ charges for the calcium (= +10), being balanced by the -1 for fluoride and the 3 x 3- for the phosphate ion = - 10).

Within these group ions, such as phosphate (PO_4^{3-}), the atoms i.e. one phosphorus atom P and four oxygen atoms O_4, have bonded together by sharing outer electrons in each atom. This is because during their formation, electrons are not easily lost but can be shared between both atoms. This is called **covalent bonding.**

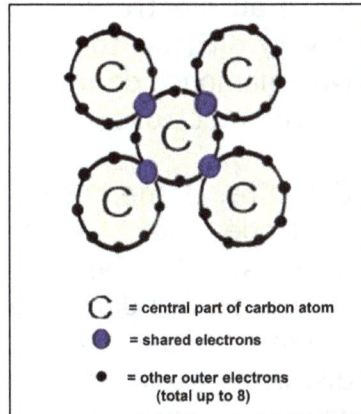

C = central part of carbon atom
● = shared electrons
• = other outer electrons
(total up to 8)

Figure 2.5: Covalent bonding as shared electrons

By sharing, these atoms then have the complete number of electrons within their outer shells to remain freely in nature and are chemically stable. For example, the natural element carbon (symbol C) has four electrons in its outer shell but needs another four more to have a stable shell of eight electrons. It can obtain this stable shell by sharing electrons with other carbon atoms.

In nature, carbon atoms come together in two forms called **allotropes** - diamond and graphite. In a diamond structure, each carbon atom has strong covalent bonds to the other carbon atoms which surround it on all sides in a strong, three-dimensional lattice.

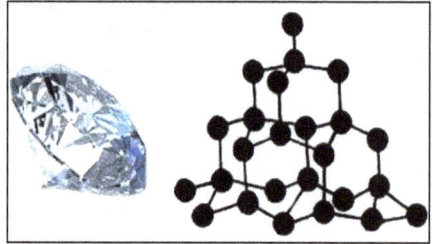

Figure 2.6: Covalent bonding in diamond. Note the strongly-packed arrangement of the carbon atoms

Since all electrons are shared within strong bonding, it is very difficult to remove electrons completely. This is why diamonds do not conduct electricity which is a flow of free electrons.

Graphite, the second allotrope of carbon, consists of sheets of covalently bonded carbon atoms separated by distances which are greater those between the atoms within the sheet. These sheets are held only by weaker

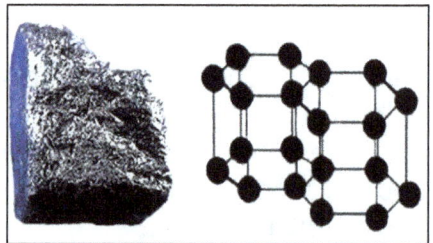

Figure 2.7: Covalent bonding in graphite. Note the arrangement of the carbon atoms in layers

forces called **Van der Waal's bonds**. This is why graphite easily cleaves in one direction and feels greasy due to sheets sliding upon one-another.

Within the sheets, the carbon atoms are strongly bonded into six-sided rings. Each set of rings has three sets of double covalent bonds as well as three sets of single covalent bonds. This arrangement is not fixed as the double bonds and the single bonds regularly switch around. This is called **resonance**. It is because of the momentary breakdown of the strong bonding during this switching which allows graphite to conduct electricity.

Figure 2.8: Resonance in carbon cyclic compounds

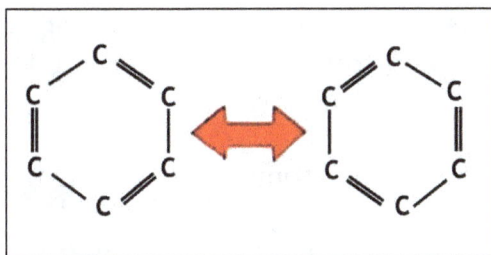

Metallic bonds are found within all metals, but especially in the metals gold, silver, platinum and copper which are the only metals to exist freely in nature. Often these native metals are found mixed together as **alloys** such as electrum, - a mixture of gold and silver with some copper.

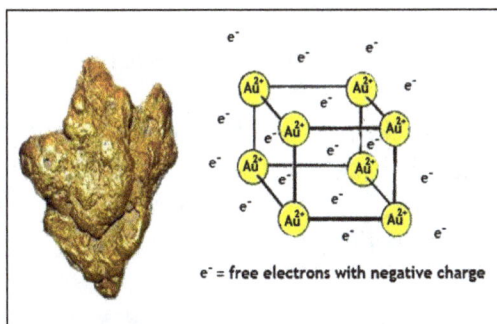

Figure 2.9: Metallic bonding in gold (Au^{2+})

Metallic bonds are similar to ionic bonds in having free electrons and electrostatic attraction as the strong force between the ions. Metals can be thought of as having a crystal lattice of ions of the particular metal packed together in a particular crystal shape and held together in a sea or glue of freely moving electrons. The total charge on the free electrons equals exactly the positive charge of the metal ions. For example, gold (with an ionic symbol of Au $^{2+}$) has a cubic crystalline structure.

Ionic substitution occurs when metal ions enter the crystal lattice of an existing mineral and replace others from this crystal lattice. How this occurs depends upon the:

- temperature

- reactivity of the ions

- size of the ions

For an ion to enter into the crystal lattice of a mineral and replace another ion, it must be similar in size. Some of the sizes of common ions are (in 10^{-10} metres) are given in Figure 2.10.

Silicon	(Si^{4+})	0.39
Aluminium	(Al^{3+})	0.57
Magnesium	(Mg^{2+})	0.78
Iron II	(Fe^{2+})	0.83
Iron III	(Fe^{3+})	0.67
Manganese	(Mn^{2+})	0.91
Sodium	(Na^+)	0.08
Calcium	(Ca^{2+})	1.06
Potassium	(K^+)	1.33

Figure 2.10: Sizes of some common ions

Some of the most important replacements include:

aluminium	(Al^{3+})	can replace silicon	(Si^{4+})
iron II	(Fe^{2+})	can replace magnesium	(Mg^{2+})
iron III	(Fe^{3+})	can replace aluminium	(Al^{3+})
calcium	(Ca^{2+})	can replace sodium	(Na^+)

The sign of the charge is not as important as the size of the ion, and substitution by ions with differing amounts of charge allows for a great variation in the chemical composition of minerals, so that some will have a specific range of chemical composition. When an ion of differing charge replaces another, the electrical neutrality of the entire lattice must be maintained i.e. the overall charge must still add to zero. This is done by adding extra ions, sometimes with different signs, into the spaces within the crystal lattice.

For example: the plagioclase feldspar series exhibits complete solid solution, in the form of substitutions, between sodium and calcium ions from albite feldspar ($NaAlSi_3O_8$) to anorthite feldspar ($CaAl_2Si_2O_8$):

Total charges for each mineral:

Albite ($NaAlSi_3O_8$)

= Na (+1) + Al(+3) + Si (+4 x 3 =+12) + O (-2 x 8=-16)
= zero

Anorthite ($CaAl2Si_2O_8$)
= Ca (+2) + Al (+3 x 2 = +6) + Si (+4 x 2 = +8) + O
(-2 x 8= -16)
= zero

Figure 2.11: Total charge for stable minerals is zero

Every atomic substitution of Na^+ by Ca^{2+} is accompanied by the replacement of a silicon ion (Si^{4+}) by an aluminium ion (Al^{3+}), thereby maintaining electrical neutrality: $Na^+ + Si^{4+} \longleftrightarrow Ca^{2+} + Al^{3+}$.

2.2 The Silicate Minerals

Silicate compounds contain the elements silicon and oxygen, with the *ate* ending refers to oxygen content in a molecule or ionic lattice. **Silicates** are the most common minerals and their great variety is due to the combining power or **valency** of the silicon atom which can attach to several other ions, the reactivity of the oxygen atom, and the ability of different ions to be substituted within a crystal lattice.

Because of the valencies of silicon and oxygen, these atoms combine to form the simplest silicate ion which is of a tetrahedral shape. In addition, these have the ability to attract to other ions or groups of ions to form more complex, three-dimensional crystal lattices.

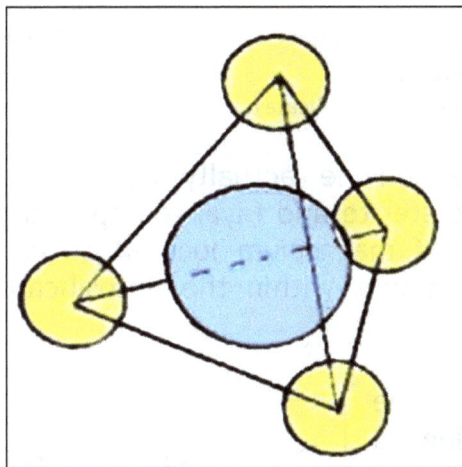

Figure 2.12: The silicate tetrahedron consists of silicon (blue) and oxygen (yellow) atoms covalently bonded but with an overall ionic charge of minus 4

A great variety of silicates is possible because the oxygen atoms can link many tetrahedra together by the covalent bonding of the oxygen atoms between them.

Ionic substitution also gives a great variety within the different families of silicate minerals. For example: in the plagioclase family, sodium ions (Na^+) can replace calcium ions (Ca^{2+}). The most important silicate mineral families are:

- **Nesosilicates** from the Greek *nesos* meaning island, these have lattices made up of separated (islands) silicate tetrahedra linked by positive metal ions.

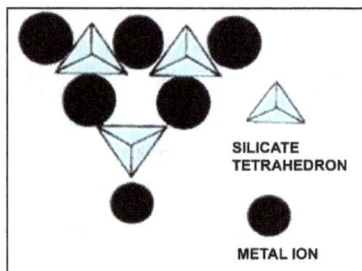

SILICATE TETRAHEDRON

METAL ION

Figure 2.13: Islands of tetrahedra

e.g. fayalite $Fe_2 SiO_4$ (iron silicate)
      forsterite $Mg_2 SiO_4$ (magnesium silicate)

The common mineral olivine actually consists of various mixtures of forsterite and fayalite depending upon the availability of magnesium ions which can substitute for the iron ions within the nesosilicate lattice.

- **Sorosilicates** from the Greek *soros* meaning a group, these consist of groups of two linked tetrahedra with metal ions within the lattice. The silicate tetrahedra are linked by a shared oxygen atom common to each.

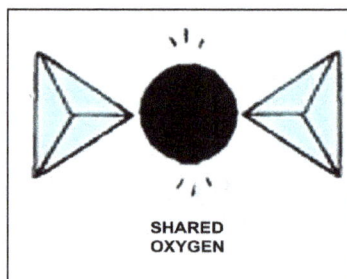

SHARED OXYGEN

Figure 2.14: Shared tetrahedra linked by an oxygen atom

e.g. melilite $Ca_2 Mg Si_2 O_7$ (calcium magnesium silicate)
      epidote $Ca_2(Al,Fe)_3 O(SiO_4)(Si_2O_7)(OH)$
                     (calcium aluminium iron silicate);
      tanzanite $Ca_2Al_3O(SiO_4) (Si_2O_7)(OH)$
                     (calcium aluminium silicate)

- **Cyclosilicates** from the Greek *cyclos* meaning a ring, these consist of multiples of rings formed by six tetrahedra linked together by sharing two common oxygen atoms. Other ions to balance charge may be found within the rings and the lattice.

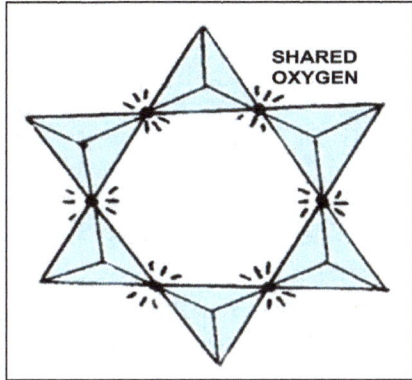

SHARED OXYGEN

Figure 2.15: A ring of tetrahedra linked by oxygen atoms

e.g. beryl/emerald
(Al$_2$ Be$_3$Si$_8$O$_{18}$) (aluminium beryllium silicate)
cordierite (Mg,Fe)$_2$Al$_3$(Si$_5$AlO$_{18}$)(magnesium iron aluminium silicate)

- **Inosilicates** from the Greek *inos* meaning a thread, these form long single or double chains of tetrahedra linked by common oxygen atoms. Other ions add around the chain(s).

They contain two major mineral groups:

1. The **pyroxenes** which have a single chain:

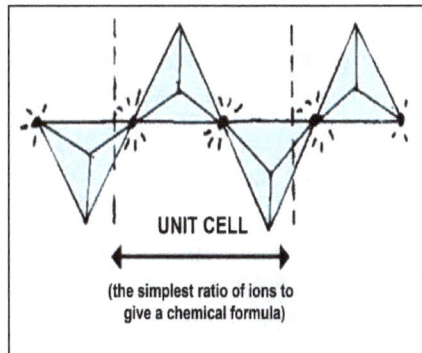

Figure 2.16: A single chain of linked tetrahedra

UNIT CELL

(the simplest ratio of ions to give a chemical formula)

e.g. enstatite $Mg_2 Si_2 O_8$ (magnesium silicate)
diopside $Ca Mg Si_2 O_8$ (calcium magnesium silicate)
augite $(Ca,Mg)(Mg,Fe,Al)(Si,Al)_2 O_8$
(calcium magnesium iron aluminium silicate)

2. The **amphiboles** which have double chains linked by an additional (i.e. a third) common oxygen atom. The holes thus formed within the structures formed by these linked tetrahedral often contain hydroxyl ions (OH-) and fluoride ions (F⁻).

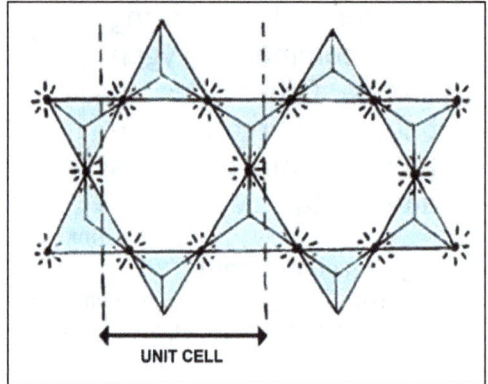

Figure 2.17: Double chain of linked tetrahedra

e.g. hornblende $(Ca, Mg, Fe, Na, Al)_{7-8} (Al,Si)_8 O_{22} (OH)_2$
(calcium, magnesium, iron, sodium, aluminium silicate hydroxide)

- **Phyllosilicates** from the Greek *phylon* meaning leaf, these have tetrahedra linked by three shared oxygen atoms giving extensive two-dimensional sheets. Within the sheet are metal ions such as aluminium (Al^{3+}), iron (Fe^{2+}) and (Mg^{2+}) magnesium, which also add to the overall bonding strength of the shared oxygen. Separate sheets are more weakly held by potassium ions (K^+).

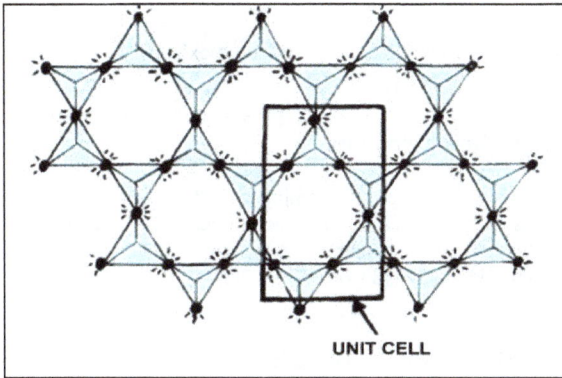

Figure 2.18: A sheet of linked tetrahedra

e.g. muscovite mica K Al_2 (AiSi$_3$) O_{10} (OH)$_2$
(potassium aluminium silicate hydroxide)
biotite mica K (MgFe)$_3$ (AlSi$_3$) O_{10} (OH,F)$_2$
(potassium, iron, magnesium, aluminium silicate
with hydroxide and/or fluoride)
talc Mg_3 (Si_4 O_{10}) (OH)$_2$ (magnesium silicate hydroxide)

- **Tectosilicates** from the Greek *tecto* meaning framework, these silicates form complex, three-dimensional frameworks in which all oxygen atoms at the apices or points of the tetrahedra are shared.

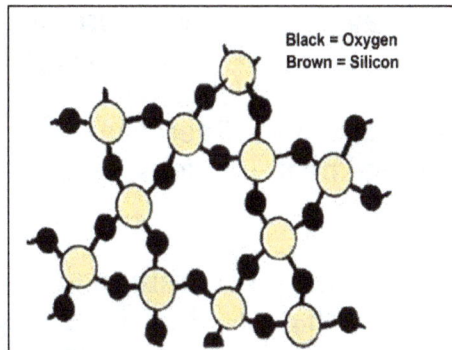

Black = Oxygen
Brown = Silicon

Figure 2.19: The 3D network of linked tetrahedra

47

e.g. quartz minerals Si O_2 (silicon dioxide)

orthoclase feldspar K $(AlSi_3)$ O_8

(potassium aluminium silicate)

plagioclase feldspars

(really a range of composition because of the substitution of calcium ions Ca $^{2+}$ for sodium ions Na $^+$ which gives a family of plagioclases ranging from high calcium aluminium silicate ,anorthite, to high sodium aluminium silicate, *albite*).

So the chemistry and the structure of minerals can be complex because of the ability of silicon atoms to bond many times, similar to the complexity of its relative, carbon in organic chemicals in living things. This chemical bonding also determines some of the observed physical properties of the silicate minerals.

2.3 More on Cleavage and Silicate Structure

Cleavage is a most useful property in identifying minerals. Recall that cleavage is the way that a mineral breaks smoothly along natural planes of weakness or flat surfaces. These planes are determined by the arrangement of ions and the forces holding them together.

Micas such as biotite and muscovite, easily split in sheets, that is, they have a perfect cleavage in one direction or basal cleavage. This is because they have strong sheets of linked silicate tetrahedra joined together by only very weak molecular (Van der Waals) bonds.

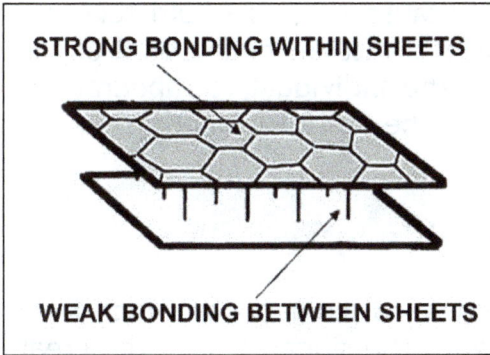

Figure 2.20: Diagram showing basal cleavage due to sheet structure

STRONG BONDING WITHIN SHEETS

WEAK BONDING BETWEEN SHEETS

Pyroxenes (e.g. augite) and amphiboles (e.g. hornblende) have perfect to good cleavage in two directions because of the weaknesses between the chains of silicate tetrahedra. Pyroxenes only have a single chain structure and so they develop a weakness giving an angle of 87° when cleaved. Amphiboles have broader, double chains and give a cleavage about 56° when cleaved.

These weaknesses and the cleavages which they produce can be best seen if the chains are observed from their ends (i.e. looking down the chains):

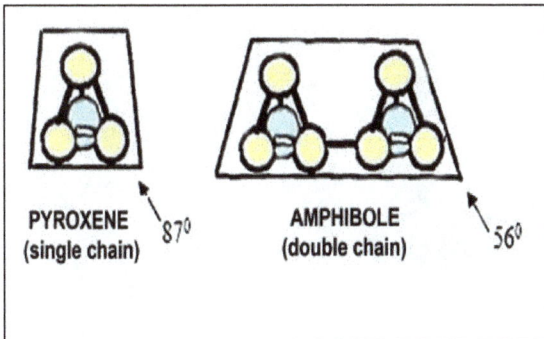

Figure 2.21: Cleavage of pyroxenes and amphiboles

PYROXENE (single chain) $87°$

AMPHIBOLE (double chain) $56°$

Olivines (nesosilicates) and quartz (a tectosilicate) do not have any cleavage because they are held together strongly in three-dimensional lattices because all of the oxygens at the apices of the individual tetrahedral are shared with the oxygens of the other tetrahedra.

2.4 Some Common Minerals

There are many other chemical compounds which form minerals. These, along with the silicates form the great majority of rocks (which are mixtures of minerals), ores (economically valuable minerals), gemstones (valuable metals and minerals used for adornment) and other earth materials. The most common groups of minerals based upon their chemistry are given in the next table:

SILICATES silicon, oxygen & other elements	OXIDES oxygen & other elements	CARBONATES carbon, oxygen & other	PHOSPHATES phosphorus oxygen & other elements	SULFATES sulfur oxygen & other	SULFIDES sulfur & other elements	HALIDES chlorine, bromine, iodine & other
Olivines $MgSO_4$ to $FeSO_4$	Quartz SiO_2	Calcite $CaCO_3$	Apatites $Ca_5(PO_4)_3$ (F,Cl,OH)	Gypsum $CaSO_4.2H_2O$	Pyrite FeS^2	Halite NaCl
Pyroxines e.g. Augite (Ca,Na) $(Mg,Fe,Al,Ti)(Si,Al)_2$ O_6	Haematite Fe_2O_3	Dolomite $CaMg(CO_3)_2$	Monazite (Ce,La,Y,Th) PO_4	Barite (Baryte) $BaSO_4$	Sphalerite ZnS	Fluorite CaF_2
Amphiboles e.g. Hornblende $Ca_2(Mg, Fe, Al)_5 (Al, Si)_8O_{22}(OH)_2$	Magnetite Fe_3O_4	Magnesite $MgCO_3$	Autunite $Ca(UO_2)_2(PO_4)_2 \cdot$ $10-12H_2O$	Anhydrite $CaSO_4$	Galena PbS	Cryolite Na_3AlF
Micas e.g. Muscovite $KAl_2(AlSi_3O_{10})(F,OH)_2$ Biotite $K(Mg,Fe)_3AlSi_3O_{10}(O$ $H)_2$	Limonite $FeO(OH) \cdot nH_2O$ Where n = any number	Siderite $FeCO_3$	Turquoise $CuAl_6(PO_4)_4(OH)$ $_8 \cdot 5H_2O$	Epsomite $MgSO_4 \cdot 7H_2O$	Chalcopyrite $CuFeS_2$	Bromargyrite AgBr
Plagioclases $NaAlSi_3O_8 - CaAl_2Si_2O_8$	Rutile TiO_2	Cerussite $PbCO_3$				
Orthoclase $KAlSi_3O_8$	Corundum Al_2O_3					
Quartz (also an oxide) SiO_2	Ice H_2O					

Table 2.1: Some common chemical groups of minerals

On the following pages are some of the most common minerals found in the field. A good knowledge of these minerals and their characteristic properties is useful in identifying them and the rocks or deposits in which they may be found. It is important to remember that these minerals will alter through reaction with natural gases and solutions (weathering) after they are exposed, so identification is often more difficult unless fresh specimens are found.

Muscovite Mica

Silver basal cleavage

Biotite Mica

Black, basal cleavage

Orthoclase Feldspar

Pink, hardness 6
2 cleavages

Plagioclase Feldspar

Cream, hardness 6
2 cleavages

Gypsum

White, waxy, hardness 2

Quartz

Glassy, hardness 7

Olivine

Olive colour, glassy,
hardness 7

Augite

Green-black, stubby
2 cleavages

Hornblende

Black, hardness 6
2 cleavages

Kaolin

White, dull, soft

Fluorite

Green, glassy, hardness 4

Halite

Clear, glassy, tastes salty

Calcite

Hardness 3
3 cleavages

Barites

Heavy,
3 cleavages

Azurite (Blue) & Malachite (Green)

Glassy to dull
hardness 3-4

Pyrite

Metallic yellow, 3 cleavages
hardness 6-6.5

Galena

Heavy, metallic, 3 cleavages
hardness 2.5

Sphalerite

Shiny, sub-metallic
hardness 4

Haematite

Black to red, dull
red streak

Magnetite

Black, magnetic
black streak

Limonite

Yellow, soft
yellow streak

Some of the major diagnostic properties of the most common minerals are given in the following table:

MINERALS	PROPERTIES									
	FORM	CLEAVAGE	FRACTURE	HARD.	COLOUR	LUSTRE	STREAK	S.G. & CHEMISTRY	CRYSTAL SYSTEM	OTHER
QUARTZ	crystals	None	conchoid.	7	varies	vitreous	white	2.65 SiO_2	Hex.	gems
ORTHOCLASE	blocks	2 @ 90^0	uneven	6	pink	vitreous	white	2.57 $KAlSi_3O_8$	Mono.	simple twins
PLAGIOCLASE	blocks	2 @ almost 90^0	uneven	6-6.5	white	vitreous	white	2.62 $NaAlSi_3O_8$ to $CaAl_2Si_2O_8$	Tri.	multi twins
MUSCOVITE MICA	Sheets	basal (1)	uneven	2-2.5	silver	vitreous	white	2.76-3.00 Silicate of Al, OH & F	Mono.	bends
BIOTITE MICA	sheets	basal (1)	uneven	2.5-3	black	vitreous	white-brown	2.80-3.10 Silicate of K, Fe, Mg, Al, OH &F	Mono	bends
HORNBLENDE	Blocks	2 @ 56^0	uneven	5-6	black	vitreous	green-brown	2.90-3.40 Silicate of Fe, Mg, Na, Al & OH	Mono	long laths
AUGITE	crystals	2 @ 90^0	uneven	5-6	black-green	vitreous	grey green	3.20-3.50 Silicate of Ca, Mg, fe, al	Mono	stubby
OLIVINE	grains	none	conchoid.	6.5-7	olive green	vitreous	white	3.20-4.40 $(FeMg)SiO_4$	Ortho.	gels with acid
CALCITE	blocks	3	even	3	white	vitreous	white	2.70 $CaCO_3$	Hex.	co_2 with acid
BARITES (BARYTE)	blocks	3	even	3-3.5	white	vitreous	white	4.30-5.00 $BaSO4$	Ortho.	heavy
KAOLINITE	massive	basal	uneven	2-2.5	white	dull	white	2.6 $Al_2Si_2O_5(OH)_4$	Tri.	earthy taste
HALITE ("salt")	blocks	3 cubic	even	2.0-2.5	white	vitreous	white	2.17 $NaCl$	Cubic	salty taste
HAEMATITE	varies	none	uneven	5.5-6.5	red	dull - metallic	red	5.26 Fe_2O_3	Trig.	red soils
MAGNETITE	blocks	3 cubic	uneven	5.5-6.6	grey-black	metallic	black	5.17-5.18 Fe_3O_4	Octahed	magnet.
LIMONITE	varies	none	uneven	4.0-5.5	yellow	dull	yellow	2.9-4.4 $FeO(OH)$ $-nH_2O$	Amorph.	Yellow pigment
PYRITE	blocks	3	uneven	6.0-6.5	gold	metallic	green-black	4.95-5.10 FeS_2	Isomet.	fool's gold
CHALCOPYRITE	blocks	indistinct	uneven	3.5	yellow	metallic	Green-black	4.1-4.3 $CuFeS_2$	Tetrag.	magnet. with heat
GALENA	blocks	3 cubic	Sub-chonchoid	2.50-2.75	lead grey	metallic	lead grey	7.2-7.6 PbS	Cubic	heavy
SPHALERITE	blocks	1	uneven	3.5-4.0	brown-black	Sub-metallic	brown	3.9-4.0 $ZnFeS$	Iso.	Shows fluores.
AZURITE & MALACHITE	varies	2	uneven	3.5-4.0	Blue & green respect.	Vitreous to dull	White – l.geen-blue	3.6-4.0 $Cu_2CO_3(OH)_2$	Mono.	blue stains in soils

Table 2.2 Properties of some of the common minerals

Chapter 3: Economic Minerals

3.1 Introduction

These are minerals which are of value and are used in the manufacture of other materials such as pure metals, alloys, fertilizers, medical products, building materials, electronics, jewellery, and are also used in power production and many other useful applications. Minerals which have economic value include:

- ores for extraction of metals e.g. haematite is the **ore** of iron

- minerals for building materials e.g. gypsum for plaster, silica (quartz) for glass, calcite (in limestone) for cement, clay for ceramics and rocks for building

- essential mineral solutions in soils for plant growth e.g. magnesium solution for green chlorophyll in leaves

- minerals as a source of raw materials for the chemical industry e.g. sulfur for making sulfuric acid, phosphate minerals for fertilizers, nitrate minerals for explosives; rock salt for chlorine production

- gemstones for decoration, industry and trade e.g. diamonds for jewelry and garnet for abrasives

Many economic minerals which are mined are **metalliferous ores**, i.e. minerals which are chemical compounds of valuable metals such as iron, lead, zinc, aluminium and uranium, and have to be processed to extract these metals. Others such as gold, silver and sulfur are found as natural elements and require minimal processing. Gems are economic minerals which are worn for adornment and are valuable because of their rarity and beauty. The mining, processing and trading of these minerals are major contributors to the economy of most countries. Some of the most common industrial minerals are given below in Table 3.1.

RESOURCE	CHEMISTRY	ORE/DEPOSIT	USES
METALS			
Iron Fe	Oxides	Haematite Fe_2O_3 Magnetite Fe_3O^4	Production of Steel
Uranium U	Oxides and Phosphates often complexes with Rare Earths and other elements	Uraninite UO_2 Davidite complex Autunite complex	Power generation, medical and Industrial Isotopes.
Gold Au	Native Element	None but may be found as mixtures/alloys	Jewellery, electronics
Silver Ag	Native Element		Jewellery, Electronics, Silverware
Platinum Pt	Native Element		Jewellery, Electronics
Copper Cu	Native Element, Sulfides, Carbonates, Oxides	Native Copper Cu Chalcopyrite $CuFeS_2$ Azurite/Malachite$Cu_2CO_3(OH)_2$ Cuprite Cu_2O	Electronics, Plumbing, Bronze (Cu + Sn), Brass (Cu + Zn) ornaments and fittings, Kitchenware, Coins
Lead Pb	Sulfides	Galena PbS Cerussite $PbCO_3$	Batteries, solder
Zinc Zn	Sulfides	Sphalerite ZnS	Batteries, Galvanising
Nickel Ni	Sulfides	Pentlandite $(Fe,Ni)_9S_8$	Batteries, Coins, Magnets, Alloys
Molybdenum Mo	Sulfides	Molybdenite FeS_2	Alloys
Manganese Mn	Oxides	Pyrolusite MnO_2	Batteries, Alloys
Chromium Cr	Oxides	Chromite $FeCr_2O_4$	Alloys, Plating, Dyes
Titanium Ti	Oxides	Rutile TiO_2	Titanium steel, White pigments
Tin Sn	Oxides	Cassiterite SnO_2	Alloys (pewter), Cans, Solder
NON-METALS			
Diamonds C	Native Element C	None	Jewellery, Cutting and Abrasives
Sulfur S	Native Element S	None - found as volcanic deposits.	Sulfuric Acid, Fertilisers, Vulcanising rubber
Phosphorus P	Phosphates (PO_4^{3-})	Apatite $Ca_5(PO_4)_3(F,Cl,OH)$ Phosphate Rock	Matches, fertilisers
Silica SiO_2	Silicates (SiO_4^{4-})	Sand and a great variety of rocks and gems	Building, Glass, Ceramics, Jewellery,
OTHER USEFUL COMPOUNDS			
Gypsum	Calcium Sulfate	Gypsum $CaSO_4$	Building, Alabaster, Plaster
Clay	Alumino-silicates	Kaolin $Al_2Si_2O_5(OH)_4$	Bricks, Pottery, Medical
Carbonates (CO_3^{2-})	Carbonates	Calcite $CaCO_3$ Magnesite $MgCO_3$	Building, Cement, Medical, Glass
Feldspars	Complex silicates	Orthoclase $KAlSi_3O_8$ Plagioclase $NaAlSi_3O_8$. $CaAl_2Si_2O_8$	Water softeners, Ceramics, Paint, Fluxes, Filters

Table 3.1: Some of the most common resources and their uses

3.2 Formation of Economic Minerals - Ores Genesis

Ore bodies around the world come in many different types and sizes, ranging from small isolated lines of ore within a rock to extensive mountain ranges made almost entirely of ore. On a small to medium scale, and of importance in the exploration by geologists in the field, is the structure known as a **gossan**. The name comes from an old Cornish miner's slang meaning blood and they are sometimes called iron hats because they are often found as raised, rounded and hard mounds. They consist of reddish oxidised products such as iron oxides/hydroxides with quartz. These represent the oxidized zone of the primary ore minerals below the surface but above the level of the water table - where the minerals can react with air and descending water to form the gossan above.

The classification of ore genesis can be complex and controversial. Without entering into such controversy, some of the main types of ore genesis are:

- **Orthomagmatic deposits** are formed from molten rock material, or magma, which has come up and intruded into the existing rock layer. These deposits are found within the igneous rocks of these large intrusions. Most of the world's nickel, chromium and platinum-group elements are derived from these deposits e.g. the Bushveld Complex of South Africa. They are often partially layered as the various minerals may have either crystallised at different times or from immiscible liquids in positions in the magma chamber and then deposited according to their density as the heavier minerals sink to the bottom. Orthomagmatic deposits also include diamonds which are found transported from great depth within volcanic pipes of **kimberlite** in which they become embedded.

- **Pegmatitic deposits** are formed from magmas which contain good percentages of dissolved water, such as those magmas which form granites and related igneous rocks. They cool at great depth, and so the crystallization process is slow and crystals of the individual minerals are large, sometimes over a metre in length. Many of these large crystalline minerals, such as feldspar, muscovite mica and quartz, can be easily quarried when the igneous intrusion is uplifted and erodes to bring it near the surface. Usually these minerals are the first to crystallize and as they tend to be minerals which do not contain water molecules in their crystal structure, an increasingly water-rich residue remains. Chemical elements such as lithium, beryllium, and niobium, which have remained in solution, become concentrated in the water-rich residual magma. At depth this residual magma may migrate elsewhere and crystallize as small bodies near the main granitic intrusion and are called rare-metal pegmatites. Pegmatites also are the major source of important gemstones, particularly tourmalines and the gem forms of beryl such as aquamarine and emerald.

- **Hydrothermal deposits** form from hot circulating water-rich fluids and are important sources of gold and the ores of mercury, antimony and copper, these are examples of sulfide compounds which are the most numerous of all ore deposits. There are several different types of hydrothermal deposits, often with a range of classification models but their common feature is that they all have been formed by the introduction of hot, metal-rich solutions from which the metals can be deposited as ores and unwanted minerals, called **gangue**.

Hydrothermal deposits are never formed from pure water, because pure water is a poor solvent of most ore minerals. Rather, they are formed from hot brine solutions such as sodium-calcium chloride solutions which also may contain traces of potassium, magnesium and some other elements. These brines are effective solvents of many sulfide and oxide mineral ores, as well as dissolving and transporting native metals such as gold and silver. The brine solutions may come from the last residual fluids from the crystallisation of magma. This hot water is expelled when some rocks are changed by heat and pressure, or from groundwater which has percolated down to a considerable depth. Regardless of the origin and initial composition of the water, the final compositions of all hydrothermal solutions tend to become similar, owing to reactions between solutions and the rocks they encounter.

Figure 3.1: Submarine hydrothermal vents near North West Eifuku Island, Japan (Photo: NOAA).

The largest deposits of this type are known as **Volcanogenic Massive Sulfide Deposits (VMS deposits)**. These have formed on the ocean floor by circulating hydrothermal fluids from volcanic vents called black smokers. Volcanogenic massive sulfide deposits are distinctive in that they are ore deposits which are associated with submarine volcanism and are independent of sedimentary processes.

Sedimentary Exhalative Deposits (SedEx deposits) are also formed by hydrothermal processes. These have been formed by release of the ore-bearing fluids into the ocean along with the deposition of the usual marine sediments. This results in the precipitation of a **stratiform ore body,** with the ores being found within the layers of the sedimentary sequence. The ore body at Broken Hill in western New South Wales, Australia is considered to be one of the world's largest silver -lead-zinc deposits. This is a complex SedEx deposit found within a metamorphic rock sequence in which the ore body has been folded and sheared. SedEx deposits worldwide are the most important source of ores of lead, zinc, silver, copper, gold, tungsten and barium metals.

Another type of hydrothermal deposit is found at the base of limestone within a marine sedimentary sequence such as that found in the Mississippi River Valley. Here, the hydrothermal solutions had penetrated the original strata, cooled and reacted with the limestone forming galena (lead sulfide) and sphalerite (zinc sulfide). This class of deposit has come to be called the **Mississippi Valley Type (MVT).**

The simplest hydrothermal deposit is a vein, a thin intrusion which forms when a hydrothermal solution flows through an open fissure and cools, with excess water boiling off or reacting with surrounding rock. This allows the crystallisation of the new ores as well as unwanted gangue minerals. A great many veins occur close to bodies of intrusive igneous rocks because the igneous rocks serve as heat sources and often provide the residual brine required for dissolving the minerals.

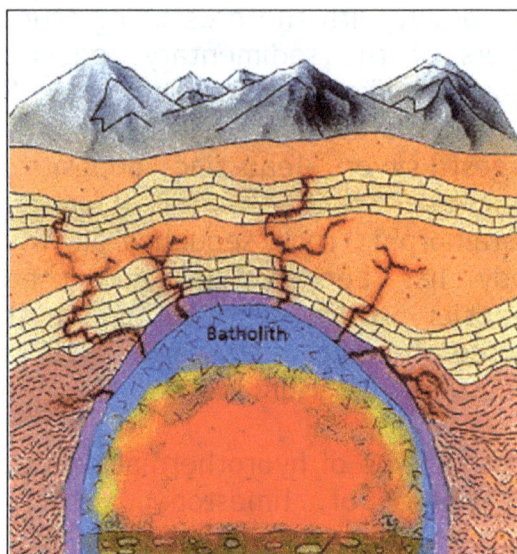

Figure 3.2: Diagram showing a batholith and associated hydrothermal veins

Other famous hydrothermal deposits include the gold-quartz veins of Kalgoorlie, Western Australia; the tin-copper-lead-zinc veins of Cornwall; Kirkland Lake, Ontario; the tin-silver veins of Llallagua and Potosí, Bolivia; and the silver-nickel-uranium veins of the Erzgebirge, Germany. The later deposits were first described by **Georgius Agricola** (German: 1494 – 1555) in his book *De re metallica* (1556), which was possibly the first reference to ores and mining techniques.

- **Banded Iron Formations (BIF)** are a most important source of iron. These are sedimentary deposits consisting of finely layered alternations of silica and iron minerals, generally haematite (iron oxide Fe_2O_3), magnetite (iron oxide Fe_3O_4), or siderite (iron carbonate $FeCO_3$). They are believed to have formed as chemical precipitates on the floors of shallow oceanic basins in highly oxidizing environments, often with some evidence of evaporation, over 2000 million years ago. Many of these deposits have been intensely deformed and metamorphosed. These deposits are extensive in the Hamersley Ranges of Western Australia, around Lake Superior in the USA, in the Serra dos Carajas of Brazil, and the Transvaal Basin of South Africa.

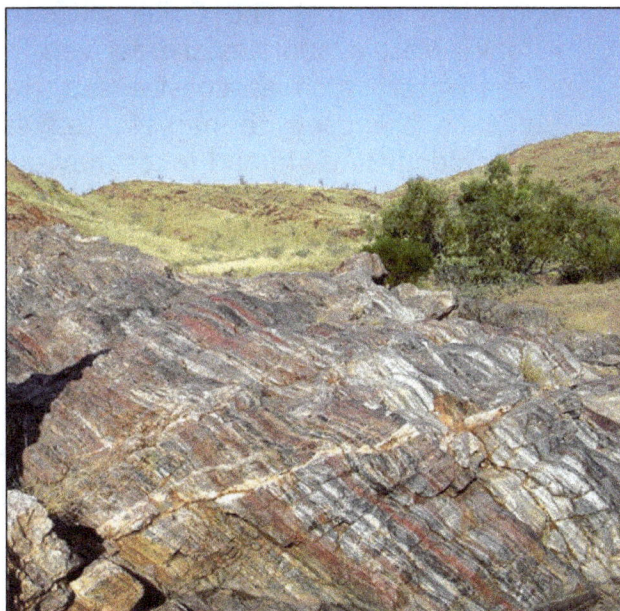

Figure 3.3: Banded iron formation, Marble Bar, Western Australia (Photo: USGS).

Deposits of manganese, an element often closely associated with iron, are also found in similar conditions. The mineral, pyrolusite MnO_4, is also found as a sea-floor deposition as manganese nodules, small balls of concentric layers of iron and magnesium mineral which are found on the sea floor near the northeast coast of Papua New Guinea and in parts of the world.

- **Evaporite deposits** are formed through the evaporation of saline and other mineral rich solutions. The most common evaporite deposits are halite (sodium chloride), gypsum (calcium sulfate), and nitrates such as saltpetre (potassium nitrate) and Chile saltpetre (sodium nitrate). Most evaporites are derived from bodies of sea-water, such as coastal salt flats, where salt pans are used to harvest sea salt. Under special conditions, inland lakes may also produce evaporite deposits, particularly in regions having volcanic springs, low rainfall and high temperature, such as the Atacama Desert in Chile and the great salt flats at Uyuni (Salar de Uyuni), Bolivia.

In places where large salt deposits have formed by evaporation then covered by sediments, they form as rock layers and are often uplifted as a salt dome. These have been mined for centuries in Europe in Poland at Wieliczka near Krakow, and in Austria at Hallein near Salzburg.

Figure 3.4: Harvesting the salts from the Salar de Uyani, Bolivia (Photo: Matthew Scott)

Online Video 3.1: Travel underground into the ancient salt mine at Hallein, Austria.
Go to https://www.youtube.com/watch?v=QCj7CjsYqHw

- **Placer deposits** are formed by gravity which takes weathered material containing the valuable minerals downhill either by water as **alluvial** deposits or as simple dry sediment slides as **colluvial** deposits. Some material may also remain at the site of weathering or nearby as **elluvial deposits**. The most common placer deposits are those of gold, platinum, gemstones, tin and heavy mineral sands containing rutile, monazite and zircon.

Figure 3.5: Black rutile in dune sand, Gold Coast, Queensland, Australia

For a mineral to be concentrated in a placer deposit, it must be resistant to weathering and erosion and have a relatively high specific gravity i.e. they are relatively heavy compared to the surrounding sediment. The placer deposits are usually found finely disseminated within the original sediments, and are only economic if they can be concentrated. This can occur by further deposition, within depressions or places where the transport medium such as a river or beach waves, have slowed down so that the heavier placer can be deposited. Such deposits include beach placers of rutile, monazite and zircon along the coast of eastern Australia, the world's main source of titanium and zirconium; the gold deposits (reefs) of the Witwatersrand of South Africa; and the placer diamonds of the coast of Namibia, southwest Africa.

- **Phosphate deposits** or lithified phosphate rock called phosphorite, is a very complex type of deposit, consisting of fine-grained mixtures of various calcium phosphates such as the mineral apatite which is a complex phosphate-hydroxide of calcium with fluorine and chlorine. Phosphorites are often formed from the reaction between marine bird excreta (guano) and the limestone rock on which the birds nest. The island of Nauru in the Pacific is one of the world's largest deposits of this type. Other sources of phosphate minerals can come from ancient, accumulated bone beds or from the leaching of phosphate-rich limestones.

Figure 3.6: Old phosphorite workings on the island of Nauru, central Pacific (Photo: Wikimedia Commons)

Chapter 4:   Mining Economic Minerals

4.1 Introduction

Apart from amateur prospecting for gold, diamonds and some gemstones, most valuable minerals must be mined on an economy of scale requiring major mining operations. The type of mine and the techniques uses in the mining and processing will depend upon such factors as:

- type of ore, especially the ore minerals available, the purity of the ore body and the gangue minerals present

- nature of the surrounding rock which can vary greatly in hardness, strength, consistency and any structural features such as faults and folds

- size of the ore body which will determine whether or not it can be mined on a massive scale or uneconomic if too small

- shape of the ore body will also determine the type of mining required; long thin veins will require a mine which will follow the ore whereas a wide, flat body would allow a more concentrated technique

- depth of the ore body may determine whether or not the ore can be mined as a surface excavation or, if deep, by shafts and tunnels

- proximity to habitation or arable land will also determine how the mine is constructed as it may be more practical to use shaft and tunnels to go beneath the existing features than to move them and mine on the surface

4.2 Stages in Establishing a Mine

There are usually several stages in establishing a mine which may require many years of preparation before the mine begins, during the life of the mine and after the extraction of the ore has ceased. The steps in the total mining process could include a sequence such as:

1. Preliminary research from previous records in government mining authority data bases, those of other companies and any local records. It is also a good idea at this early stage to find out information about local land-ownership, especially any traditional tribal affiliations or national park/conservation issues.

2. Application and obtaining any permits required by government. These may include exploration permits, access to government or tribal lands, and any relevant environmental permits.

3. Exploration to locate the resource by geologists and geophysicists, often using detailed exploration techniques, both surface and sub-surface, until the ore body or resource is located. Whilst it is not a requirement to involve local land-owners at this stage, it is usually good etiquette to discuss the exploration with them. Often land-owners can be most helpful in pointing out access or land features.

4. Estimation of the extent and value of the ore body using the field data and 3-dimensional computer modeling. In addition there will be a chemical assay of field samples to determine the grade of the ore or resource considering the purity of the ore by analysing how much waste mineral or gangue, and rock is mixed with it. These two factors will give some prediction as to the value and viability of the future mining operation.

5. Conduct mine planning to evaluate the economically recoverable portion of the deposit and giving some initial information needed for the future construction of the mine site.

6. Conduct a feasibility study to evaluate the total project and make a decision as whether to develop or walk away from the project. This includes a complete analysis of the possible mine and its site from the initial environmental impact study, through the total mining and recovery process, transportation and markets, and finally

the closing of the mine and subsequent land reclamation and restoration of the habitat.

7. Formal application and obtaining appropriate government mining leases, environmental permissions, royalty agreements with landowners, both legal and traditional, and access rights prior to starting the operation.

8. Development of the site, including all mining operations, transportation routes with access to and from the site, soil storage, waste dumps and environmental nurseries, accommodation facilities with communal infrastructure and on-site processing facilities if required.

9. Extraction and processing of the ore on a large scale with on-going environmental testing and application, especially monitoring the water table and any leakage of harmful solutions from the mine, and then shipping to ports or to distant factories for further processing.

10. Reclamation to make land once more suitable for future use either as farmland or urban development. Many mining companies employ professional environmentalists who will oversee the filling in of pits and shafts then replacement of topsoil. After this, native vegetation will be replanted from local seedlings obtained before the mining started, and native animals will be encouraged to return. In some instances, the

mining company will re-sculpture the land for the specific uses of the landowner or local township.

4.3 Introduction to Mining

Mining can be considered as either surface mining or underground mining. Which type is used depends upon the:

- nature and shape of the ore body

- depth of the ore body

- location of the ore body in relation to habitation, land-use and places of importance

- potential costs, as a mine will often begin as a surface operation but as costs of removing the unwanted rock from the pit becomes prohibitive; the operation may switch to underground mining

4.4 Surface Mining

Surface mining is the predominant excavation method procedure worldwide. It involves the removal of surface material such as vegetation and top soil and then either the large-scale excavation of a pit, which may shallow or deep depending upon the nature and depth of the ore, or other such as sluicing, dredging or drilling and extraction to extract the ores. The most common forms of surface mining include:

- **Open pit**, also called open cut or open cast mining, is usually employed to exploit a near-surface deposit or one that has a low stripping ratio of unwanted rock or **overburden**, to ore. **Benches** in the pit describe the horizontal levels of the mine and act as roads. They are usually on 3 metre or 6 metre levels, depending on the size of the machinery being used. Most walls of the pit are generally dug on an angle less than vertical, to prevent and minimise damage and danger from rock falls. The inclined section of the wall is known as the **batter,** and the flat part of the step is known as the bench or **berm**. The size of the bench and the angle of inclination of its walls depend upon the rock type and its degree of weathering. Any structural weaknesses which may occur within the rocks must be stabilized first. These weaknesses include: joints or large cracks; faults where joints have moved; and foliation or weak folding planes within the rock itself. In some instances additional ground support is required and rock bolts and shotcrete (a mesh bolted to the rock face and then sprayed with concrete) may be inserted into the bench walls. De-watering bores may be used to relieve water pressure within the rock walls by drilling horizontally into the wall and allowing trapped water to flow into well-constructed sumps and drains for pumping out of the pit.

Figure 4.1: Gold-copper open pit mine, Cadia Hill, central New South Wales, Australia,

Figure 4.2: Detail of a bench drilled for explosive charges (right).

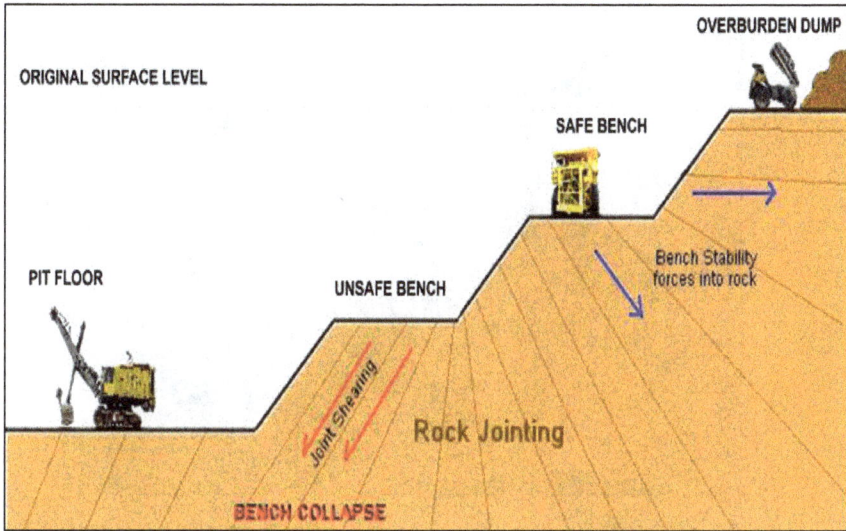

ORIGINAL SURFACE LEVEL

OVERBURDEN DUMP

SAFE BENCH

PIT FLOOR

UNSAFE BENCH

Bench Stability forces into rock

Joint Shearing

Rock Jointing

BENCH COLLAPSE

Figure 4.3: Diagram showing an open cut pit with benches. Note the concept of stability of benches in relation to local rock jointing. If shearing (sliding motion) occurs along the joint planes, then the bench will collapse.

A haul road is situated at the side of the pit, forming a spiralling ramp, up which heavy haulage trucks can carry ore and waste rock as well as to allow the movement of other plant such as drill rigs, excavators, bucket excavators, dozers, water tankers and personnel vehicles to new areas of drilling, blasting and excavation.

Figure 4.4: Drilling rigs drill a pattern of holes in the bench for inserting explosive charges

Figure 4.5: A bench drilled for explosives. These boreholes will be filled with liquid explosives and then primed with detonators by the shot-firers (professional explosives experts)

Figure 4.6: Bucket excavators are used to shovel the broken material into trucks.

Figure 4.7: The large bucket of an excavator

Figure 4.8: Excavator loading a haulage truck

Figure 4.9 Dozers rework mined debris and help to construct roads

Figure 4.10: Mine haulage trucks can carry loads from 40 to 400 short tons

Figure 4.11: Empty mine haulage truck travelling down the bench road to the work face.

Figure 4.12: Graders help make the haulage roads

Figure 4.13: Water tankers spray the roads to reduce dust

The original topsoil and the waste rock from the pit are piled up at the surface near the edge of the open cut in waste dumps. These waste dumps are also tiered and stepped, to minimise erosion and are monitored by the mine environmentalist for any leaching of toxic solutions which may run off into the local catchment area.

Waste material left over from the initial processing of the ore is mixed with water and removed as tailings. This is removed as a slurry of fine ore particles in water. This is then pumped into a tailings dam or settling pond, where the water evaporates and the solids are removed or buried. **Tailings** dams often contain toxic substances due to the presence of heavy metal compounds in the gangue and chemicals which have been used during the processing of the ore e.g. cyanide which is used to treat gold.

Figure 4.14: A tailings dam

After the mine has been closed, the mine area is then rehabilitated. Sometimes the initial stage of this conservation process occurs during the mining operation, with old sections of the pit being back-filled with waste rock as mining proceeds elsewhere in the pit. When the mining operation is complete, the filled pit and waste dumps, including the stored topsoil, are contoured to flatten and stabilise the land surface and vegetation is replanted. If the ore contains heavy metals, sulfides and other toxic substances, it is initially covered with a layer of clay to exclude air and surface water which can oxidise the sulfides to produce corrosive sulfuric acid.

In old pits which have not been reclaimed, there eventually will be some erosion and further damage to the environment, with a considerable time being needed until the dumps become acid neutral. These dumps are usually fenced off to prevent livestock becoming poisoned and also from denuding any vegetation that might otherwise stabilise the ground. Large pits may also fill with ground water with the potential for further hazards by contamination and drowning.

Figure 4.15: Mt. Lyell, Tasmania, Australia. An old copper mining region which has never been fully rehabilitated.

Figure 4.16: A younger disused pit at Mt. Lyell, Tasmania

- **Strip mining** is usually done over a wider area but at shallow depth, often with only one bench. It is used for deposits such as bauxite, coal, buried alluvial gems, ancient mineral sands and some building material. Heavy equipment such as graders and dozers remove the overburden and then mine the deposit strip-by-strip, back-filling as each section has been mined. For larger deposits, huge dragline excavators are used. Smaller deposits, especially building materials, such as the rocks limestone, sandstone and slate, usually are quarried in smaller, compact sites.

Figure 4.17: Strip mining of heavy mineral sands from ancient beach sands (now inland) at Iluka, southwestern Victoria, Australia

- **Placer mining** is used to exploit loosely consolidated deposits such as common sand and gravel or gravels containing gold, tin, diamonds, platinum and titanium. This may involve:

  - **Hydraulicking** utilizes a high-pressure stream of water which is directed against the mineral deposit to remove it by the erosive action of the water. This is especially effecting in extracting tin in Malaysia and gold from old tailing deposits in South Africa.

Figure 4.18: Using high pressure water used to sluice surface deposit (Photo: Wiki Commons)

- **Dredging** is carried out from floating vessels either on natural waterways over the ore or on man-made ponds over mineral-rich sands. For example, manganese nodules as balls of concentric layers of iron and manganese oxides are mined at sea off the north east coast of Papua-New Guinea using large grab dredges on ships. In some of the sand islands off the coast of eastern Australia, mineral sands, a mixture of the minerals rutile, zircon and ilmenite in sand, are mined using suction or bucket line dredges. These scoop up the sand which is then transferred to a concentrator which separates the heavy mineral sands from the lighter quartz sand using washed agitated screens. The heavy minerals remain on the screens, and the lighter quartz sands are expelled out of the rear of the dredge where it can be reworked and stabilized with vegetation.

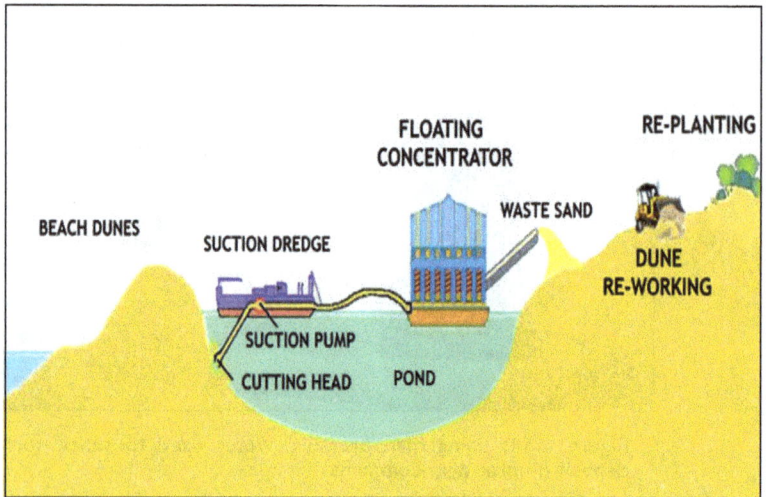

Figure 4.19: Diagram of a suction Dredge in a mineral sands mining operation

Figure 4.20: Bucket-line gold dredge in Alaska scoops up the gold-bearing sediment and separates the gold on a grooved, vibrating sluicing table. (Photo: USGS)

- **Panning** is an old technique which is often used in small scale and individual separating of valuable materials such as gold. The technique relies upon the relative heft, or heaviness of the wanted material compared to the soil or gravel in which it is contained. The scale of the operation can be increased using large, industrial, vibrating and grooved separators, flushed with water to carry away the lighter soil and leaving the heavier metal behind.

Online Video 4.2: Watch a demonstration of gold panning technique.
Go to https://www.youtube.com/watch?v=B5h7H4nG7aM

Figure 4.21: The Hand of Faith gold nugget discovered in 1980 near the surface in Victoria, Australia, using a metal detector.

Online Video 4.3: Travel back in time to Sovereign Hill, Victoria, Australia - a re-enactment gold rush town, mine and watch a gold smelting operation.
Go to https://www.youtube.com/watch?v=xLQUb9tarNA

- **Borehole Mining (BHM)** involves the drilling of boreholes into a subsurface ore body and its extraction using high pressure water jets or steam. It can be used in a variety of environments such as on the land surface, within existing open pit floors, within underground mines or from floating vessels using pre-drilled boreholes. It is often used when the subsurface ore can be dissolved in high pressure (and often hot) water, steam or chemical solutions. An

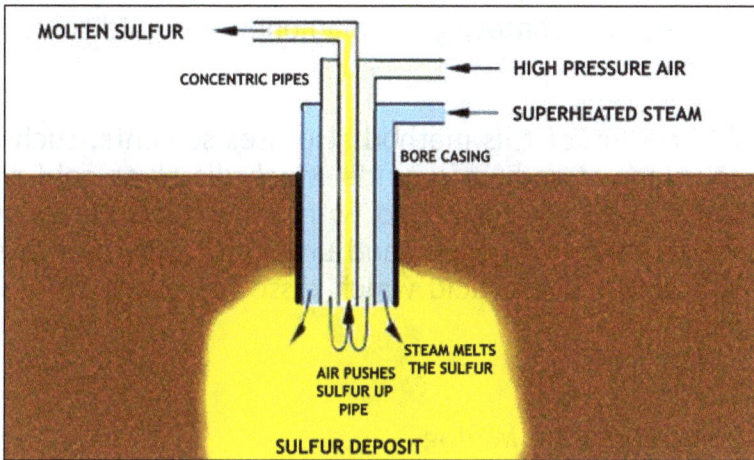

Figure 4.22: The Frasch Process uses BHM to extract sulfur

example of the use of this technique is in the extraction of underground salt (sodium chloride) or sulfur. Initially a borehole is drilled into the ore body and a casing pipe is lowered into the hole followed by the BHM tool. This tool consists of two concentric pipes – the outer pipe carries the high pressure solution down, and the inner pipe brings the dissolved or fragmented ore to the surface as a slurry. This is then processed, the ore extracted and the waste water re-used. The lower end of this tool may also contain a pump to maintain the liquid pressure and sometimes a small cutting head to assist with the cutting of the ore body. Borehole mining usually has a relatively low capital cost, ease of mobility, a relatively small surface footprint with low environmental impact and the ability to work in hazardous and dangerous conditions. This method has been used in mining of uranium, quartz sand and gravel, phosphates, gold, diamonds and gold to name a few.

A variation of this method also uses solvents, such as solutions of sodium cyanide which dissolves gold and silver, weak sulfuric acid with ammonium carbonate which dissolves copper and uranium from their ores, and hydrochloric acid which dissolves boron from its salts.

4.5 Underground Mining.

Underground mining usually occurs when the:

- Ore body is deep, rendering open pit unviable because of the large amounts of overburden which must be extracted and therefore the huge size of the surface pit.

- Ore body lies beneath habitation centres or valuable arable land and it would be too expensive to give up the surface for mining.

- Ore body is not compact and is a complex of veins or smaller bodies which are best mined by following the shape of the ore rather than taking out a considerable amount of waste.

- Ore body is vertical or nearly so, and it can be mined using a primary shaft with secondary drives or tunnels.

Typically the ore body is reached by either vertical shafts or **declines,** which are sloping tunnels, often in huge spirals which go down into the ore; or **adits** which are horizontal or gently sloped tunnels which enter from the sides of mountains or existing open pits. The entrances to declines and adits are usually strongly supported by iron and concrete **portals** preventing any blockage from collapsing **country rock.** Levels within the mine may also be connected by small vertical shafts called **winzes.**

Figure 4.23: The old shaft elevator at Mt. Lyell copper mine, Tasmania, Australia

Figure 4.24: Modern entrance to the main decline of the Mt. Lyell copper mine, Tasmania, Australia

Online Video 4.4: Join a group of student geologists descending to an upper level of a gold mine at Ballarat, Victoria, Australia
Go to https://www.youtube.com/watch?v=k240TTg9F60

Inside, the mine may be either supported if the surrounding rock or the country rock is weak and the ore body is fragmentary, or unsupported if the country rock and ore have considerable strength. If the mine is of the supported type, the ore is mined section by section, often being allowed to collapse under its own weight, a process called **caving**. Then the fractured ore is mined from the collapsed and partly opened ore body called a **stope**. Stope excavation can be from the base as overhand stoping, allowing the ore to descend in a controlled manner, or from the top as underhand stoping. Sometimes when the ore body is nearly horizontal, the stope may be mined along horizontal access tunnels with different parts of the ore body being blasted vertically so that only one section of ore at a time would collapsed and then be mined, often at several levels. This is called sublevel stoping.

A variety of equipment such as excavators, ore skips and rail cars, drills and loaders is used below ground. Often the larger pieces of equipment are constructed within the mine and then left below as part of the backfill at the end of its usefulness. In dangerous stopes, the heavy equipment can be operated by remote control from a console further down the tunnel or from the surface.

Figure 3.25: Pneumatic drill used for drilling the rock face

Figure 3.26: A rock face drilled and partially set with explosive charges

Figure 4.27: An older mining skip used to haul ore in mines

Figure 4.28: A mine haulage truck entering a small portal of an adit

Localised bracing and strengthening of the walls and roof may be done using ring bolts with an epoxy glue to stabilise fractures, static supports or moveable hydraulic supports. As the stope collapses, the space above becomes jammed with large slabs of country rock and so there would be little, if any subsidence on the surface.

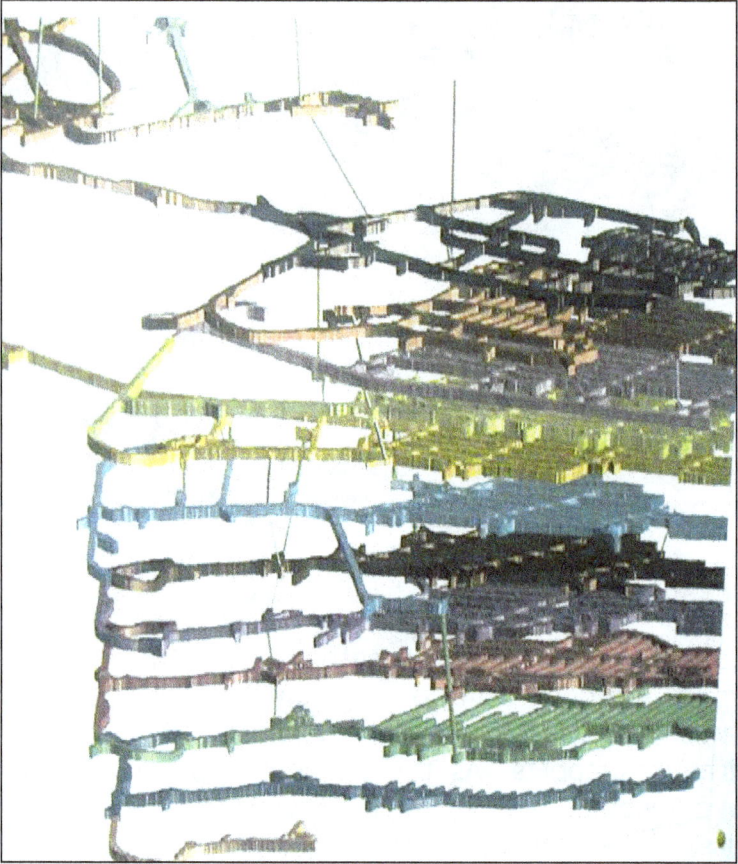

Figure 4.30: 3-Dimensional map of a subsurface working showing the maze of tunnels of the Mt. Lyell Copper Mine, Tasmania, Australia.

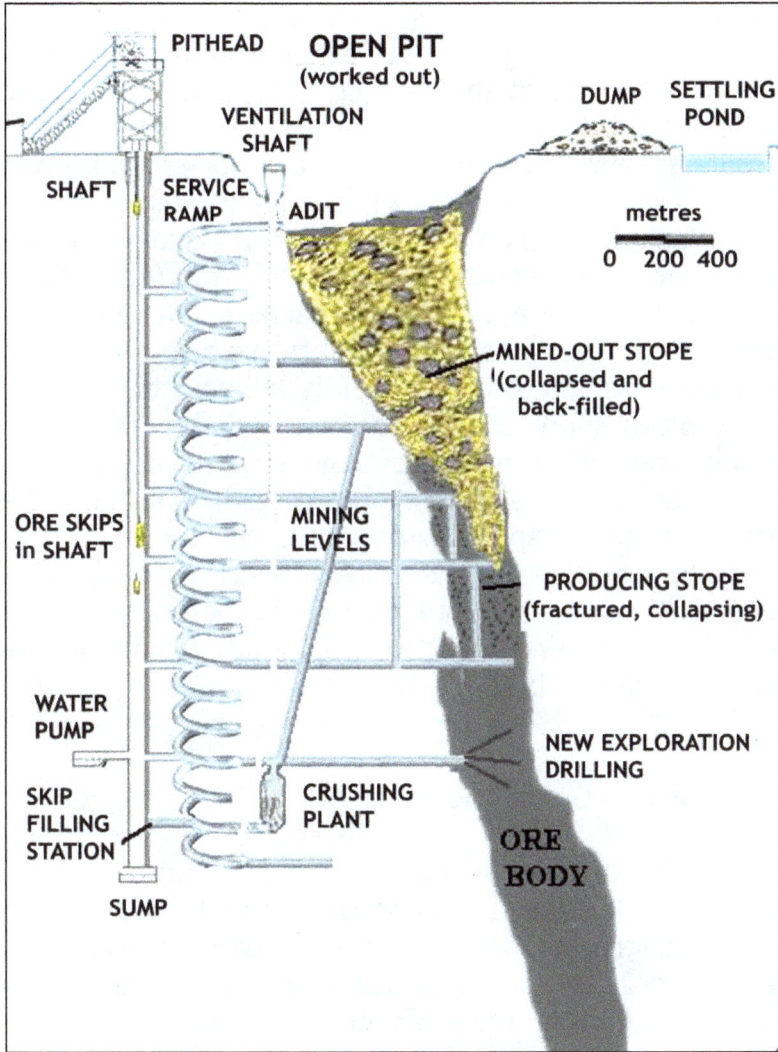

Figure 4.29: Diagram showing a simplified stope (collapsing ore body) mining operation

If the ore body is tabular and horizontal or nearly so, and the county rock is structurally strong, then unsupported mining would be the most efficient technique to use. In the mining of coal, limestone and salt, the roof is generally supported by natural columns or pillars, and the mine is excavated around them leaving tunnels or **bords**, or even very large spaces called rooms. Supplementary ring bolts and mesh are also added in places where the roof is fractured. When the end of the ore has been reached, the mining is reversed in direction and the pillars are also mined in a controlled fashion and the roof is allowed to cave in. Miners often shore up the roof with timber poles and walls of the collapsing areas or **goafs**, with dry timbers which crack at the slightest failure, giving warning of a possible collapse and allowing time for the miners to leave. In many cases the mined areas are also back filled with waste and old machinery before they are collapsed. Further details of coal and its mining are given later in this book.

4.6 Gems and Gemstones

A **gem** is anything which is valued and worn for adornment. It could be made of any material, such as black-lacquered wood, red coral, plastic, glass beads, feathers or bones. **Gemstones** are gems made from naturally-occurring minerals or rocks. Gemstones are valued because they are often rare, have a clear diaphaneity and lack inclusions or bubbles of gas or water, and usually have a hardness of 6 and above on Mohs' Scale of hardness. Gemstones are usually graded by the 4Cs – Colour, Clarity, Cut and Carat Weight (1 **carat** weight = 0.2 g. with the name being derived from the ancient weight of a carob seed used in trading).

The demand or popularity of a particular gem depends upon scarcity and current fashion. Scarcity or the availability of a gem is controlled by the limitations of its natural abundance, the politics of its country of origin, and the abuse of monopolies. The last factor is a dominant feature of very valuable gems, such as diamond. Mining companies may be grouped together as a monopoly and control the export and trade of their gems in order to manipulate the prices or put any competition out of business.

Gemstones can be classified in many ways, but a common way is by their value into precious and semi-precious gemstones. There are also rarities and ornamental stones. Rarities are minerals which are extremely difficult to obtain because they are not common in nature, such as the rare, peach-coloured version of beryl called morganite. Ornamental stones are those which have been carved or kept as natural crystals and are used as household decoration such as pieces of fossilised wood and carved jade.

- **Precious Gemstones** are relatively rare and usually have very high value, usually over $5000 per carat, and include:

  - Tanzanite is blue-purple calcium-aluminium hydroxyl silicate.

Figure 4.31: A single crystal of tanzanite and a faceted (or cut) gem of tanzanite

- Diamond, from the ancient Greek *adámas* for unbreakable, is crystallised carbon, formed at within the Earth's mantle over 100 km below the surface. The iron-magnesium and carbon rich magma is then brought near the surface in the pipes of explosive **kimberlite** volcanoes. Kimberlite volcanoes are found in South Africa and are named after the city of Kimberley. They are also found in the United States and Russia. Synthetic diamonds can also be made industrially by subjecting graphite to extremes of temperature (over 1500 ^{0C}) and pressure (5000 million pascals).

Figure 4.32: a natural diamond crystal and a faceted stone

| ROUND or BRILLIANT | PEAR | OVAL | MARQUISE |
| HEART | SQUARE | EMERALD | RADIANT |

Figure 4.33: Some of the common cuts of diamond made by cleaving and grinding the crystal with diamond dust

- Ruby is a form of corundum or aluminium oxide (Al_2O_3). It is usually a deep red colour due to the distorted crystal structure of the aluminium oxide lattice and sometimes the additional replacement of some of the aluminium ion (Al^{3+}) with that of chromium (Cr^{3+}). Its name comes from *rubrum*, the Latin for red. On Mohs' Scale, it has a hardness of 9.0. The early main source of natural rubies was from Burma (now Myanmar) but they also found in Thailand, Cambodia, Afghanistan, India and Australia to name a few.

Figure 4.34: Natural ruby crystals – note their hexagonal crystal shape – and a crystal cut as an oval

- Sapphire is another form of corundum – aluminium oxide. It is usually blue in colour, but can also be found in nature as yellow, purple, orange, green and sometimes a mixture of colours as "parti sapphires". The colours are due to trace amounts of other elements such as iron, titanium, chromium, copper or magnesium. Padparadscha sapphires are a pink-orange colour and are very rare. They are found in Sri Lanka, Vietnam and East Africa. Productive sapphire deposits are also found in Australia, Thailand, China, Madagascar, and in North America. A star sapphire shows a star-like pattern of radiating light known as **asterism** when view in good light from above. Some sapphires contain intersecting needle-like inclusions (water or air bubbles) which follow the crystal structure causes the appearance of the six-rayed star-shaped if the crystal is cut in the correct orientation as a rounded stone or **cabochon**. The name is an ancient one and has been used as a personal name in both Hebrew (*sappir*) and Greek (*sappheiros*) cultures for this red gem.

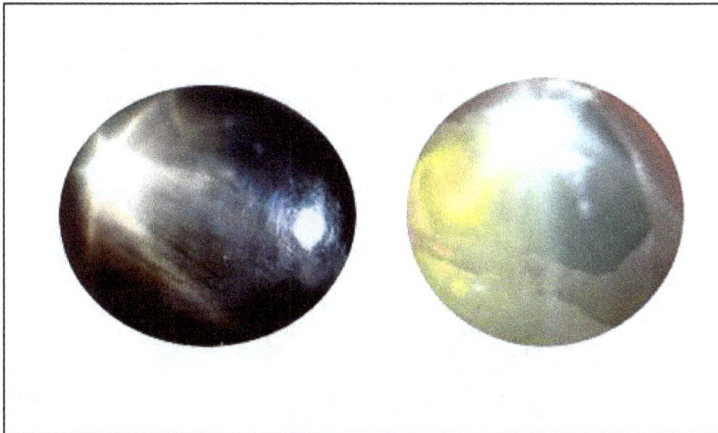

Figure 4.35: Star sapphires: traditional (left) and green (right) (Photos: P. Lynch-Harlow)

- Emerald is a form of green beryl, a cyclosilicate of the element beryllium ($Be_3Al_2(SiO_3)_6$. The name was originally derived from the Ancient Greek *smaragdos* for green gem. Beryls are usually a cream-yellow colour, but also include the pale blue gem, aquamarine (Latin: *aqua marina* for water of the sea) and the pink variety, morganite, named after the industrialist, J.P. Morgan. Chrysoberyl is another name for several beryllium gemstones which include yellow-green alexandrite and cymophane (or cat's eye).

Figures 4.36 & 4.37: Natural emeralds in weathered Igneous rock from Brazil (left) and cut emeralds from Torrington New South Wales, Australia (photo at right: P. Lynch-Harlow)

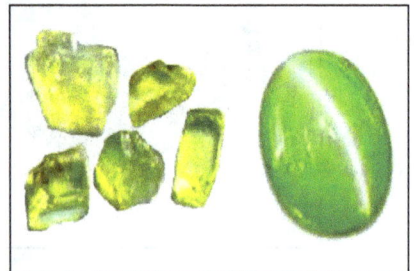

Figures 4.38 & 4.39: Natural crystals of aquamarine (left) and of cymophane (right) with a cabochon or rounded gemstone cut to give the cat's eye effect when light is shone directly from above it

Emeralds are mined in Columbia which produces most of the world's stones. Zambia is the next main source of the gem but other sources include Egypt, Austria, Afghanistan, Australia, India and parts of eastern Africa.

- Black opal is hydrated silica (silicon dioxide with water – $SiO_2.n\,H_2O$, where n = any number) with the water molecules within the crystal lattice. It consists of microscopic spheres of pure silica (silicon dioxide) regularly arranged. These spheres refract and disperse light, breaking up into a rainbow spectrum, giving the play of colours seen in this gemstone. The opal may not be exactly black but more likely a variety of colours, but the darker stones are the most favoured. In poor quality opal, or potch opal, the spheres are irregularly arranged and so that the light is poorly dispersed and the colours are often not visible.

Figure 4.40: North Queensland boulder opal as rich veins of opal within a dull, brown ironstone.

Figure 4.41: A collection of black opals – note the great variety of colour (Photo: P. Lynch-Harlow)

- **Semiprecious gemstones** are the more common gemstones and are usually (but not always) of lower value. They include:

  - Quartz Gemstones made from quartz (silicon dioxide) with impurities which give their different colours. All have a hardness of 7 on Mohs' Scale of hardness and include:

    - Amethyst is purple in colour and named from the Ancient Greek terms: *a-* and *méthystos* for not and intoxicated. It was believed that this gemstone could prevent drunkenness and so it was used as a charm worn on the body or as a powder put into drink. The purple colour is due to natural radiation and traces of iron impurities and some trace elements. Artificial irradiation can be used to further enhance the colour.

Figure 4.42: An aggregate of amethyst crystals

- Citrine is yellow, and its name comes from the Latin: *citrina* for yellow due to the impurity of iron III (Fe^{3+}). It comes mostly from Brazil.

Figure 4.43: A cluster of citrine crystals crystals

- Smoky quartz is usually grey and translucent and is also called cairngorm from the mountains of that name in Scotland. The colour is due to aluminium-based and irradiation-induced coloured centres.

Figure 4.44: Smokey quartz crystals

- Rose quartz in which the pink colour is due to trace amounts titanium, iron, or manganese;

Figure 4.45: A massive piece of rose quartz

- Rock crystal is usually clear, transparent and is sometimes cut as imitation diamond but it is a lot softer with only a hardness of 7 compared to 10 for diamond on Mohs' Scale. It sometimes contains inclusions such as golden hair-like crystals of the mineral rutile (titanium dioxide) as rutilated quartz.

Figure 4.46: An aggregate of rock crystal (left) and a polished crystal containing fine threads of rutile (right)

- Agate is a cryptocrystalline form of quartz called chalcedony which has very fine crystals which can only be seen under a microscope. Agate has attractive banding and examples include:

  chrysoprase (green chalcedony)

  onyx is black chalcedony

  sardonyx and jasper are both red

Chalcedony is precipitated from hot, quartz-rich solutions into cavities such as veins and vesicles or gas holes, in recently-cooled lavas. The oval or spherical vesicles may be partly or completely filled with fine crystals of quartz or more solid chalcedony and are then termed **amygdales**. Sometimes the surrounding igneous rock, such as a rhyolite, may weather more quickly and become a clay soil, leaving the hard, resistant quartz shapes as **geodes** (or thunder eggs).

Figure 4.47: A chalcedony geode

Figure 4.48: Agate is banded chalcedony

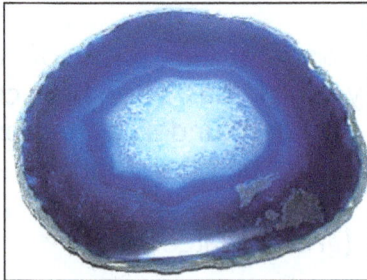

Figure 4.49: A thin slice of dyed Brazilian agate is very translucent

Figure 4.50: A geode (thunder egg) of agate which has been dyed blue. It is from Brazil where the geodes are placed in a fire pit to open micro-cracks, then strong dyes are added which are absorbed into the crystal

Figure 4.51: A cabochon of red, banded agate (left) and uncut piece of massive chrysoprase (right)

- Garnets are red, yellow, green and blue cubic nesosilicates and the name comes from the Middle English *gernet* for dark red, which is the most common colour of this stone. Their hardness is from 6.5 to 7.5 and garnet is often found as small grains in garnet sands and these grains are useful as an abrasive. Larger, better-formed and clear specimens are excellent gemstones.

Figure 4.52: Natural crystals of red almandine garnets. Note the natural cubic crystal of the large stone

Figure 4.53: A mandarin garnet – a spessartine garnet from Tanzania, Africa - faceted and mounted with diamonds (Photo: P. Lynch-Harlow).

- Peridot is olive-green in colour and is the gem-quality of the common mineral, olivine - iron-magnesium silicate (Mg^{2+} - Fe^{2+})$_2$ SiO_4 although the gem variety has more magnesium than iron. It is often found in vesicles within basalts and other iron-magnesium rich igneous rocks.

Figure 4.54: Uncut peridot crystals – the gem variety of the common mineral olivine

- Topaz can be various colours, but it is often clear or as the more valued blue topaz – both are forms of aluminium silicate with fluoride and hydroxide ions - $Al_2SiO_4(F,OH)_2$. It is named after a legendary island Topazos in the Red Sea which was also difficult to find in legendary Arabian sea-faring lore. It is often found associated with quartz-rich granitic rocks in such places as Russia, Afghanistan, Sri Lanka, parts of Europe, Pakistan, Japan, Australia, the Americas and Africa.

Figure 4.55: Pieces of rough topaz showing basal cleavage and two faceted stones; a teardrop (left) and brilliant cut of blue topaz (right)

- Turquoise is blue-green hydrated copper-aluminium phosphate and its name comes from the French: *turques* for Turks. It is a waxy, cryptocrystalline, opaque gemstone much used in Ancient Egypt and Mesoamerica.

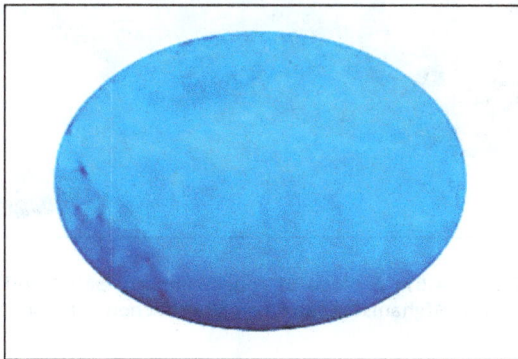

Figure 4.56: A turquoise cabochon

- Lapis Lazuli is metamorphic rock which is deep blue and often mottled with veins and spots of golden pyrite, white quartz and calcite. The name comes from several languages meaning blue stone. It has been used in many ancient cultures as gemstones and as inlay work in costumes and buildings. It has been mined from ancient times in the Sar-i Sang region in Afghanistan, especially in the Kokcha Valley, the legendary site of the Lapis mine of the Queen of Sheba. Today, it is also mined in Russia, Afghanistan and the Andes Mountains.

Figure 4.57: A piece of rough, impure lapis lazuli from the Kokcha Valley, Afghanistan, and a small cabochon cut from it.

- Moonstone is sodium-potassium feldspar - $(Na,K)AlSi_3O_8$ - and the diffuse reflection from the stone is due to successive layers of the feldspar within the cabochon. It is found in Armenia, Myanmar, Austria, the United States and Australia.

Figure4.58: A moonstone cabochon

- Iolite is the gem variety of the mineral cordierite, a magnesium iron aluminium cyclosilicate found within metamorphic rocks. The name iolite comes from the Greek word *ios* for violet. This mineral has strong **pleochroism**, i.e. it will change colour at different angles of light when the cut stone is rotated. The pleochroism of iolite is from blue to yellow grey. It also has the effect of polarising light and may have been used by the Vikings as a *sólarsteinn* or sunstone to locate the position of the Sun after sunset or within a sea fog.

Figure 4.59: Iolite in the emerald cut shape (Photo: P. Lynch-Harlow)

4.7 Making Gemstones

Some gems are synthetics, which are man-made substances which imitate, and are identical in chemical composition and optical quality to an original rock or mineral gemstone. For example, diamonds, emeralds, rubies and sapphires can be grown in the laboratory by subjecting powder chemically identical to the original material, to great heat and pressure and then slowly cooled. Other materials which look like an original gemstone but do not have the same chemistry as the original can also be fashioned into gems. These are called imitations. For example, imitation diamonds can be made from glass, quartz and cubic zirconia.

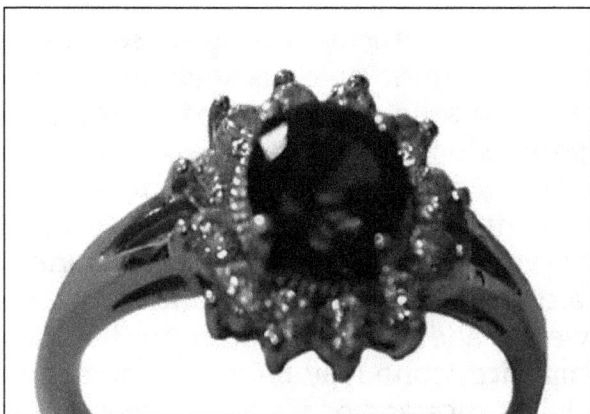

Figure 4.60: A cheap ring-setting with imitation ruby and diamonds - all made of cubic zirconia

Often gemstones may be enhanced to improve their appearance by:

- heating such as to darken amethyst, sapphires, rubies and amber

- bleaching to whiten pearls and jade

- dying to change the colours of agate and some pearls

- irradiating with gamma rays to enhance and darken blue topaz

- coating of tanzanite, diamond, topaz and coral to enhance or change their colour to more attractive tints

Opal can also be darkened to enhance the natural colours by boiling them in a sugar solution and then treating with sulfuric acid. In addition, because of the expense of the solid gemstone, many are manufactured as composite stones. In this process, a very thin layer of the original gemstone is cemented to slices of a less valuable material such as quartz, clear topaz or the synthetic version of the gem using colourless cement which also has good optical qualities. By doing this, added weight and strength is given to the original slice. The **composite stone** may be in the form of a doublet, or an original slice and one layer of cheaper material, or as a triplet which has an original slice and two layers, one of which may be a coloured backing to enhance the overall colour. Black opals are often sold as doublets or triplets, but so are other gemstones such as sapphires, emeralds and diamonds. With luck, the buyer can see the joins using a good hand lens.

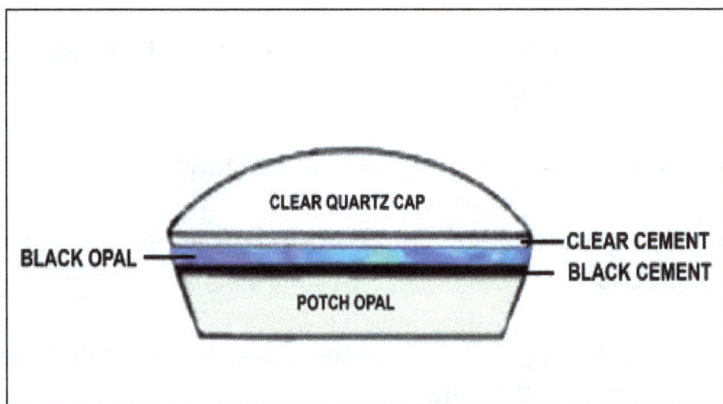

CLEAR QUARTZ CAP

BLACK OPAL — CLEAR CEMENT / BLACK CEMENT

POTCH OPAL

Figure 4.61: A triplet of black opal which is actually a composite stone of a very thin layer of solid black opal cemented with black cement to cheaper potch opal to enhance the dark colour and to give bulk. These layers are then capped by cementing a rounded layer of transparent quartz to enhance the overall reflectivity of the triplet.

Gemstones are shaped by a long process of cleaving, **faceting** and polishing. The cut or shape of the gemstone is also a major factor in its value along with colour, clarity and carat (weight). In ancient times, gemstones, especially in Europe, were cut and ground in a free-hand style into any nice shape, or simply as a rounded cabochon with a flat bottom, rounded top and cut in an oval or circular shape. The Frenchman **Jean Baptiste Tavernier** (1605-1689), returned to Europe after his travels in India, and introduced the art of faceting stones into various shapes with many flat faces or facets. This is done by grinding and polishing the stones with many flat faces at different angles. These facets increased the external and internal reflectivity of the stones and gave a more pleasing effect. Gemstones have been a major industry in Asia, especially in China, India and Southeast Asian countries well before the European trade. Many of these places are still some of the World's major cutting centres.

Figure 4.62: A faceting wheel and attachments (Photo: P. Lynch-Harlow)

In Europe, the diamond-cutting industry is centred in Antwerp in Belgium and Amsterdam in the Netherlands. Major cutting centres also located in Tel Aviv in Israel and in Dubai in the Middle East. Diamonds are not cut in the literal sense, but are cleaved by placing a thin blade along a perceived cleavage line. Diamonds have a cubic crystal with three cleavages and once the blade is positioned, it is lightly tapped so that the rough diamond is split along the cleavage plane. The rough shape may then be sawed using a diamond saw. This is a metal wheel which has small, industrial diamonds around its upper edge. The diamonds may also be **brunted** or rounded, using other diamonds, into the approximate shape intended for the stone. The pieces can then be finished by faceting using a wheel coated in diamond dust and then polished.

Figure 4.63: A master cutter faceting a diamond on a wheel lightly covered with oil and diamond dust

Lapidary is the skill of cutting gemstones and is a common hobby as well as a trade in many countries. Many towns and cities have lapidary clubs for people to learn these skills at a modest cost especially in countries where gemstones can be located by weekend prospecting.

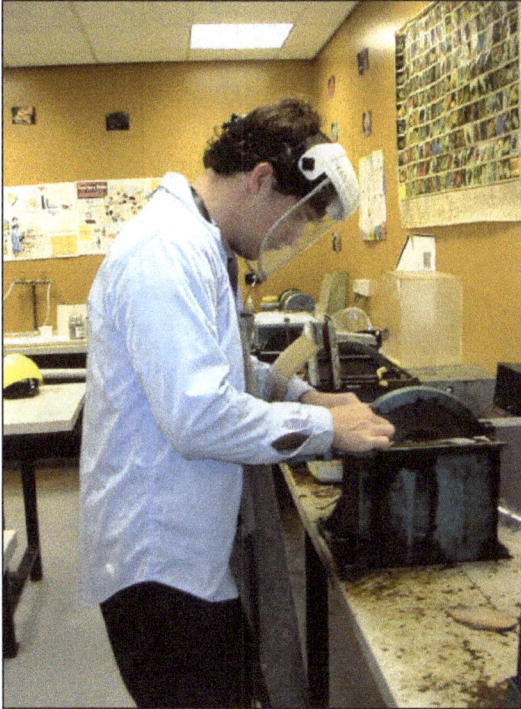

Figure 4.64: In a lapidary laboratory – using a diamond-studded saw to cut a slab of agate.

Chapter 5: Processing the Mined Ore

5.1 Introduction

Mining the ore is just the first phase of a long series of processes designed to extract the final resource, especially metals from the Earth. Other resources such as building materials, coal, sulfur and salt may involve only simple processing, but obtaining pure metals often involve very complex refining processes. This usually involves:

- crushing of the mined material

- separating the ore from rock and gangue

- converting the ore to a usable, concentrated product (often an oxide)

- smelting the concentrate to the final product (e.g. a metal)

5.2 Crushing the Ore

Crushing is usually done within the mine or immediately on the surface using stamper batteries using piston hammers, jaw crushers using opposable teeth, or large rolling mills which use large balls of iron in rotating drums to crush the material. After crushing, the material is then screened for size and then passed on so that the valuable ores can be separated from unwanted minerals or gangue, and rock.

Figure 5.1: Two rolling mills used on site to crush ore

Figure 5.2: Iron balls used in rolling mills to crush ore

5.3 Separating the Ore

Separating uses processes which remove the unwanted rock, sediment and the gangue from the crushed and screened ores. Sometimes there may be several economic ores mixed together, for example, lead and zinc minerals such as galena and sphalerite respectively, as these are often found together along with gold and silver and gangue minerals of quartz and feldspars. Separation techniques are usually carried out at the mine site in a processing plant, and the separation techniques usually make use of a difference in property between the wanted ore and the unwanted rock and gangue. Some of these techniques include:

- Density differences, with the denser ore being allowed to sink as the lighter, unwanted material is washed away. Often the material is agitated across a sloping grooved table such as in the separation of gold or heavy mineral sands.

Figure 5.3: Wet agitator benches which separate heavy mineral sands the lighter quartz sands which continue with the water

- Centripetal Force is the outward directed force experienced when objects are spun around in a circle. When a mixture of heavier ore and lighter gangue is passed down a spiral-separator or **cyclone** in a water stream, the heavier ore particles in the rotating stream have too much inertia to follow the tight curve of the stream, and so they stay on one side of the spiral and then fall to the bottom where they are removed. Lighter sands and rock are pushed to the outside of the curve and are removed to one side at the bottom of the spiral.

Figure 5.4: Cyclones removing light sand from heavy mineral sands. As the water, ore and soil is spun around the spiral channel, the lighter materials are flung out to the outer edge

- Adherence and buoyancy conditions are produced within a mixture of water, powdered ore and gangue which is agitated with an air stream in **froth flotation** separators. This technique can selectively separate different useful minerals by adding a flotation agent, such as sodium ethyl xanthate (or SEX - $CH_3CH_2OCS_2Na$), which has both hydrophobic or water hating, and hydrophilic or water loving, properties. Added to the bubble mixture, the hydrophobic part of the xanthate molecule attaches to the ore particles which are then buoyed up to the top of the tank as bubbles formed by the agitated air. Alkaline solutions of about 10% and pH of 7-11 (NB: pH is a measure of the acidity of a solution, with 7-14 being alkaline and 0-7 being acid) are normally used. By adjusting conditions of temperature, air flow, and amount of chemical, the process can be used to separate individual powered minerals such as sphalerite (zinc sulfide) and galena (lead sulfide). The resulting froth can then be removed and settled as a concentrate of the target mineral by running the separated froth into settling tanks so that the ore can settle from the slurry. It is then dried and stored.

Figure 5.5: froth flotation vats showing the powdered ore being separated by adhering to agitated bubbles produced using air, water and chemicals.

Figure 5.6: A settling tank where the processed ore is allowed to fall to the bottom and water and chemicals are removed.

Figure 5.7: Removed from the settling tanks, the ore is dried to a fine powder

Figure 5.8: The final purified mineral is stock-piled as a fine, dry powder ready for smelting

- Electrostatic separation involves the electrical charging of the ore mixture using a positively-charged electrode and a negatively-charged drum on which the mixture falls. Many factory chimneys have electrostatic collectors which collect smoke particles. Minerals which can be charged electrostatically adhere to the drum for a movement and then fall off separately to minerals which are not electrostatic. For example, heavy mineral sands contain the minerals rutile, ilmenite, zircon and monazite which can be used in this process. Once crushed to a fine powder, the rutile and ilmenite are charged and then separated from the other two.

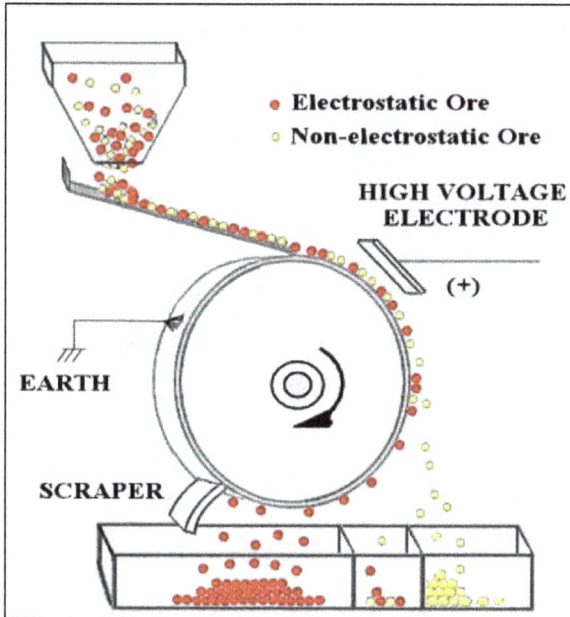

Figure 5.9: Diagram showing separation by electrostatics

- Magnetic separation can be used to separate ores which differ in magnetic properties using a conveyor belt passing over a magnetised drum. Magnetic ore, such as ilmenite and monazite can be separated from the non-magnetic ore. Using a combination of electrostatic and magnetic separation, a heavy mineral sands mixture can be separated into its component minerals. Magnetic separation is also useful in separating magnetic magnetite from other iron oxides such as limonite and haematite.

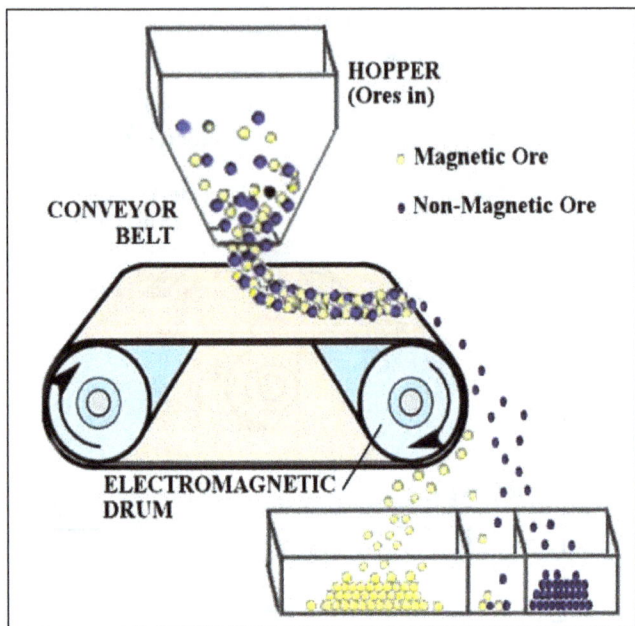

Figure 5.10: Diagram showing separation using electromagnetism

Figure 5.11: A small magnetic separator

- Solubility differences between some minerals or their gangue make separation easy when one mineral is soluble. The Celts in central Europe over a thousand years ago, channelled water into the salt mines at Hallein, Austria, dissolving the salt which was then channelled outside into evaporation pans and then crystallised (see Evaporites previously in this book).

Gold is soluble in sodium cyanide (NaCN) solution and forms a soluble complex of gold cyanate. This is useful in mining gold which is of a low concentration within a rock or sediment and which cannot be seen by eye nor easily dredged. Crushed rock, such as gold-bearing granite, or alluvial sediment is piled into large mounds irrigated with sodium cyanide solution. Over about a week, the runoff will flow into a dam which has been isolated from the environment using plastic sheeting liners because the cyanide solutions are highly toxic. From here, the solution is filtered and passed over powered zinc which is more reactive as a metal than gold. The zinc replaces the less active gold from its solution producing gold-plated zinc. This is then smelted in a furnace and the heavier molten gold sinks to the bottom where it can be tapped off into ingots of good purity. The remaining molten zinc can then be re-used in the process.

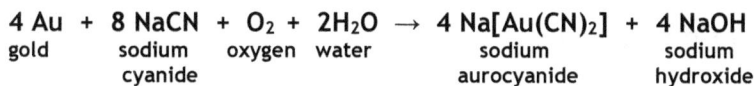

$$4\ Au\ +\ 8\ NaCN\ +\ O_2\ +\ 2H_2O\ \rightarrow\ 4\ Na[Au(CN)_2]\ +\ 4\ NaOH$$

gold   sodium   oxygen   water   sodium   sodium
        cyanide   aurocyanide   hydroxide

When zinc dust is added it is replaced by the gold:

$$[Au(CN)_2]^- + Zn \rightarrow 2Au + Zn(CN)_4^{-2}$$

aurocyanide   zinc   gold   zinc cyanide
ion   solution

This is a gross over-simplification of this process, as there are other metals such as copper also mixed with the gold which also dissolve in the cyanide solution. In addition, all of the solutions are toxic to the environment and waste disposal is difficult.

- Bacterial action has been used more recently to extract gold mixed with unwanted sulfide minerals. Bacteria, such as *Acidithiobacillus ferrooxidans*, are used to oxidise the sulfide minerals which trap fine gold, converting the sulfides (S^{2-}) to soluble sulfates (SO_4^{2-}), releasing the gold which then can then be recovered using the cyanide process. This process has also been used in the extraction of copper.

Bacteria have also been used within the cyanide process to extract the gold using treated resins and biopolymers, renewable organic molecules of long chain structure such as cellulose, grown from natural fibres using specialised micro-organisms. The latter process is far less toxic, makes use of simple, natural materials and is renewable to some extent.

5.4 Converting the Ore

Conversion to a more concentrated and pure chemical compound, such as an oxide or sulfide, is the final step before the pure metal can be extracted by smelting. This process may be done on site but if a more complex process is required, the mine product may be shipped to a refining plant elsewhere.

For example, iron concentrate from a mine is probably in the form of powdered oxides as haematite Fe_2O_3 or magnetite Fe_3O_4. Copper concentrate is usually a sulfide from the minerals chalcopyrite $CuFeS_2$, chalcocite Cu_2S, covellite CuS and bornite $2Cu_2S \cdot CuS \cdot FeS$ and it also contains some iron, gold and silver. Copper carbonates, such as malachite/azurite $CuCO_3 \cdot Cu(OH)_2$ are easily converted to oxides by roasting in air or dissolved in sulfuric acid.

$$CuFeS_2 \ + \ 2O_2 \ \rightarrow \ FeO \ + \ CuO \ + \ SO_2$$
chalcopyrite   oxygen   iron   copper   sulfur
                          oxide   oxide   dioxide gas

$$CuCO_3 \cdot Cu(OH)_2 \ + \ 2H_2SO_4 \ \rightarrow \ 2CuSO4 \ + \ CO_2 \ + \ 3 H_2O$$
malachite/ azurite   sulfuric acid   copper sulfate   carbon   water
                                   solution   dioxide gas

Aluminium ore, bauxite, a complex mixture of aluminium oxide (Al_2O_3) and aluminium hydroxide ($Al(OH)_3$) undergoes a series of conversion processes involving crushing, dissolving in concentrated sodium hydroxide ($NaOH$) solution and then heating at a high temperature to get pure, white aluminium oxide called alumina. This is explained further in this chapter.

5.5 Smelting the Ore

Smelting is the final extraction of the pure metal from the ore or impure state. Only gold, silver, platinum and copper can be found in nature as native metals or as their alloys. Gold and silver have been known to humankind and prized for their beauty and use in adornment. Platinum is more difficult to find and to work, and copper is more commonly found as its chemical compounds as sulfides, carbonates and oxides. Iron, aluminium and other metals are more chemically reactive and form a great variety of chemical compounds. These metals have to be separated from the other chemical elements in their compounds by processing and smelting. This often involves complex processes to remove the other elements, especially oxygen which is very reactive, from the required metal. Some of the major smelting processes involve:

- **Copper Smelting** - copper, as bronze which is a mixture or alloy with tin, was possibly the first metal to be smelted and fashioned by humankind. The name copper probably came from the Latin name *aes cyprium*, or metal of Cyprus, where it was first mined. This term became abbreviated to cuprum and hence the symbol for copper as Cu. The bronze formed from copper and tin which in ancient times in Europe often came from the tin mines in Cornwall, United Kingdom, was much harder than pure copper. Brass, the alloy of copper and zinc, was also used by the Romans for household items and decoration.

In the Flash Process developed by Outokumpo in Finland in 1949, the dry, powdered, copper concentrate is usually an impure mixture of copper sulfides (such as the minerals chalcopyrite $CuFeS_2$ and chalcocite Cu_2S), iron sulfides, arsenic and antimony compounds, and small amounts of gold and silver. This is added to powdered limestone and silica as sand, and blasted with a stream of oxygen-enriched air within a gas-fired furnace. The process occurs within the stream of gas and fine concentrate and goes through a series of complex reactions which can be simplified to:

$$2CuFeS_2 + O_2 \rightarrow Cu_2S + 2FeS + SO_2$$

Chalcopyrite   oxygen   copper I   iron   sulfur
(copper iron sulfide)   sulfide   sulfide   dioxide gas

The iron component, now as iron sulfide (FeS), is oxidised to iron oxide (FeO) by the oxygen in the gas stream.

$$FeS + O_2 \rightarrow FeO + SO_2$$

iron   oxygen   iron II   sulfur
sulfide   oxide   dioxide gas

The copper component, mainly copper I sulfide ($Cu2S$) falls to the floor of the furnace as **copper matte**. The silicon dioxide in the sand reacts with the limestone, the iron oxide and other impurities to form a **slag**, mainly as a mixture of iron and calcium silicates ($FeSiO_3$ and $CaSiO_3$). This floats on top of the matte and can be run off. Once the reaction has started, it produces its own heat as an exothermic reaction which then maintains the reaction. The waste sulfur dioxide is removed to the acid plant where it is converted to sulfuric acid.

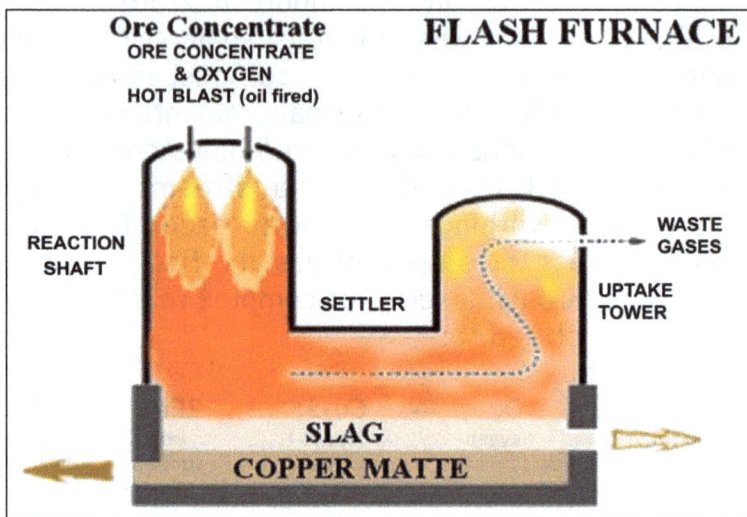

Figure 5.12: Diagram showing the Flash Process

The copper matte is then blasted with oxygen in a converter furnace to produce impure (98%) blister copper.

$$2\ Cu_2S \quad + \quad O_2 \quad \longrightarrow \quad 4\ Cu \quad + \quad SO_2$$

copper I   oxygen   copper   sulfur dioxide
sulfide   gas

The **blister copper** is then roasted in an anode furnace to produce **anode copper** (about 99% pure) suitable for the next processing step.

Electro-refining is the final step in which the copper as plates, as anodes or positively-charged electrical terminals, are immersed in an electrolytic solution of copper sulfate and sulfuric acid in a bank of electrolytic cells. An electric current is passed through the solution, and copper from the anode is carried through the electrolyte as copper ions Cu $^{2+}$, and deposited in a pure form on the negatively-charged cathode starting sheet made of pure copper. Impurities from the anode, such as lead, zinc, nickel, arsenic, selenium, tellurium, gold and silver, fall to the bottom of the electrolytic cell as a slime which can be further processed to extract any of the other valuable metals. This refining produces copper of more than 99.9% purity.

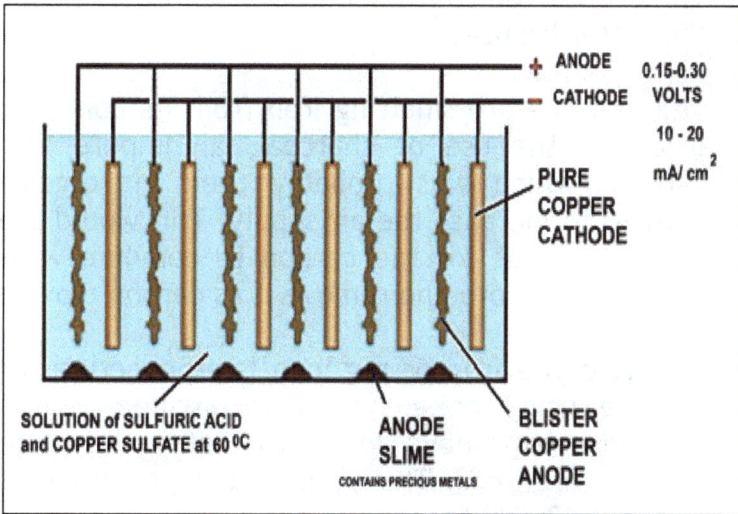

Figure 5.13: Diagram showing the electro-refining of copper

- **Iron Smelting** - Iron has the chemical symbol Fe – from the Latin *ferrum,* and is one of the most abundant elements in the Earth. It is not found in the native state because it is moderately reactive and readily forms compounds such as the oxides haematite (Fe_2O_3), magnetite (Fe_3O_4) and limonite ($FeO(OH).nH_2O$) as well as many others. Rust and most of the red discolouration in rocks and soils are due to these compounds.

Iron requires a higher temperature (about 1500 ^0C) to melt than bronze (about 950 ^0C), so it was not until about 1200 BC that humankind, firstly in Anatolia, in Turkey, started to fashion iron tools rather than using the softer bronze.

The main key to smelting iron from its common iron oxides is the use of charcoal, an impure form of carbon, which was probably used in forges with bellows to increase the air supply. This would provide sufficient heat for the carbon to combine with the oxygen in the ore and remove it as carbon dioxide.

Today, crude iron or **pig iron** is produced in a blast furnace using **coke**, which is distilled coal, with limestone to remove any silica that would be mixed with the iron ore. This is done at high temperatures of about 1200^0C for six to eight hours.

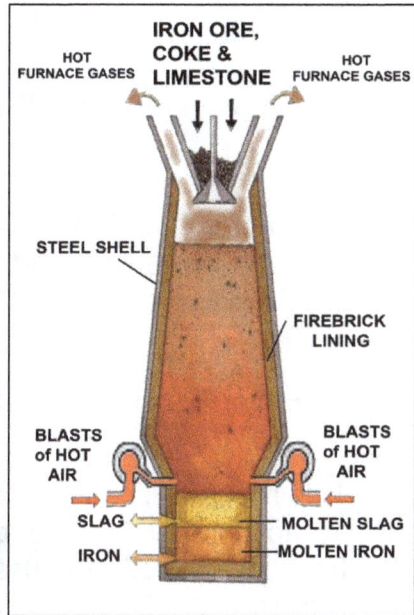

Figure 5.14 & 5.15: Blast furnace at Port Kembla, New South Wales, Australia (left), and a simplified diagram of a blast furnace.

The coke descends to the bottom of the furnace to the level where the preheated air enters the blast furnace. Here the coke is ignited, reacting immediately to produce heat and carbon dioxide:

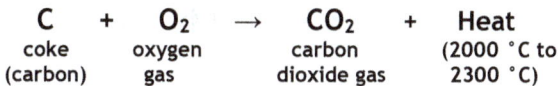

$$C \;+\; O_2 \;\rightarrow\; CO_2 \;+\; \text{Heat}$$

| coke | oxygen | carbon | (2000 °C to |
| (carbon) | gas | dioxide gas | 2300 °C) |

Since the reaction takes place in the presence of excess carbon at a high temperature the carbon dioxide is then reduced to carbon monoxide:

$$CO_2 \quad + \quad C \quad \rightarrow \quad 2CO$$

carbon	carbon dioxide gas	carbon monoxide gas

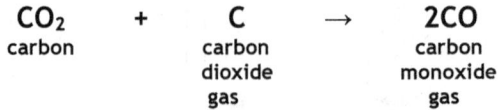

Limestone is added and descends in the blast furnace and remains as a solid while going through its first reaction:

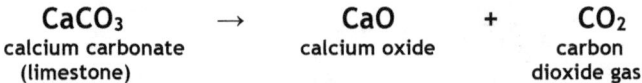

$$CaCO_3 \quad \rightarrow \quad CaO \quad + \quad CO_2$$

calcium carbonate (limestone)	calcium oxide	carbon dioxide gas

This reaction requires considerable energy and starts at about 870°C. The calcium oxide (CaO) formed from this reaction is used to remove any sulfur from the iron which might contain iron sulfide impurities. The waste produced from the resulting with the calcium sulfide (CaS) is removed as a slag:

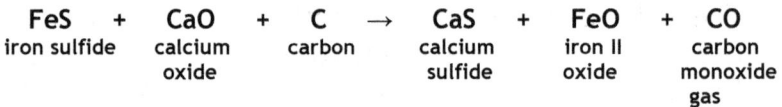

$$FeS \; + \; CaO \; + \; C \; \rightarrow \; CaS \; + \; FeO \; + \; CO$$

iron sulfide	calcium oxide	carbon	calcium sulfide	iron II oxide	carbon monoxide gas

The calcium oxide also removes some of the silica (SiO), which is the rock/sand in the mixture, in the slag:

$$CaO \quad + \quad SiO_2 \quad \rightarrow \quad CaSiO_3$$

calcium oxide	silicon dioxide	calcium silicate

142

The iron ore then has its oxygen content removed (i.e. is reduced) by a series of chemical reactions using the carbon monoxide previously produced:

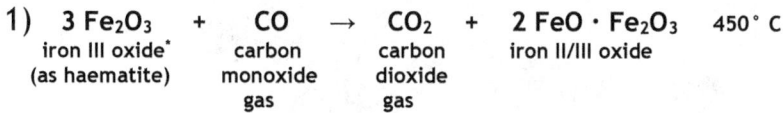

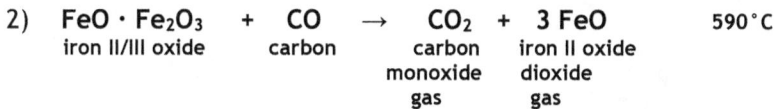

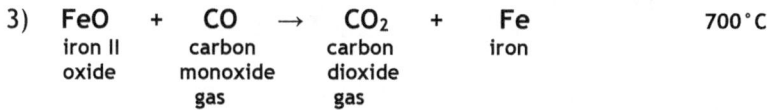

1) $3\ Fe_2O_3$ + CO → CO_2 + $2\ FeO \cdot Fe_2O_3$ 450° C
   iron III oxide*   carbon   carbon   iron II/III oxide
   (as haematite)   monoxide   dioxide
   gas   gas

2) $FeO \cdot Fe_2O_3$ + CO → CO_2 + $3\ FeO$ 590°C
   iron II/III oxide   carbon   carbon   iron II oxide
   monoxide   dioxide
   gas   gas

3) FeO + CO → CO_2 + Fe 700°C
   iron II   carbon   carbon   iron
   oxide   monoxide   dioxide
   gas   gas

(* Note: iron, like many other chemical elements has more than one reaction ability - or **valence**. Iron can be iron II when it loses 2 electrons from its outer atomic shell in reactions or iron III when it loses 3 electrons. As electrons are negatively charged, this gives the iron ion -charged atom - valencies of Fe^{2+} and Fe^{3+} respectively)

Figure 5.16: Torpedo-shaped ladle rail cars which take the molten pig iron to other parts of the steel works for further processing.

The pig iron, so named because the early iron ingots reminded the workers of the shape of pigs, is cast into moulds for various items as **cast iron**. This, however is very brittle and easily broken because there is still some carbon within the iron as an impurity. To remove this, the iron must be further processed into **steel** by oxidising the carbon to carbon dioxide.

There are two major commercial processes for making steel:

1. **Basic Oxygen Steelmaking (BOS)**, which uses liquid pig-iron from the blast furnace and scrap steel, and accounts for over 60% of the world's steel production. In this method, a **flux** of dolomite (calcium magnesium carbonate) is added to the mixture to assist in the reaction, and then a blast of pure oxygen is applied to remove the carbon from the iron. The heat comes mainly from the reaction itself, which is exothermic i.e. the reaction generates extra heat.

The vast majority of the world's steel is manufactured using the basic oxygen furnace accounting for over 60% of global steel output. Modern furnaces will take a charge of iron of up to 400 tons and convert it into steel in less than 40 minutes, compared to 10-12 hours in the old open hearth furnace.

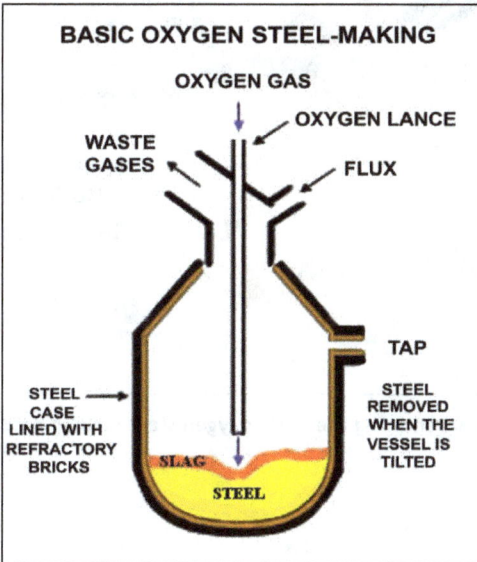

BASIC OXYGEN STEEL-MAKING

OXYGEN GAS

OXYGEN LANCE

WASTE GASES

FLUX

TAP

STEEL CASE LINED WITH REFRACTORY BRICKS

STEEL REMOVED WHEN THE VESSEL IS TILTED

SLAG

STEEL

Figure 5.17: Diagram showing the basic oxygen steel-making process

2. **Electric Arc Furnace (EAF)** steelmaking, which uses scrap steel or iron already reduced by natural gas as the main feed materials. Large amounts of high voltage electricity are used to melt the solid iron and oxygen is blown into the melt to remove any carbon. Lime (calcium oxide CaO) and fluorspar (calcium fluoride CaF_2) are added as fluxes to assist the melting process at about 1800^0C. A mid-sized modern steelmaking furnace would have a transformer rated about 60 MVA, with a secondary voltage between 400 and 900 volts and a secondary current in excess of 44,000 amperes. To produce steel in an EAF requires approximately 440 kWh per metric tonne.

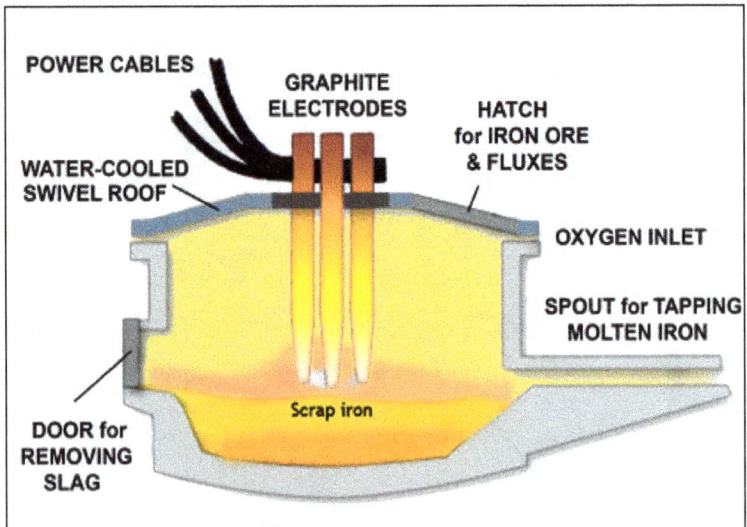

Figure 5.18: Diagram showing the basic oxygen steel-making process

- **Aluminium Smelting** - of the modern industrial metals, aluminium was the last to be refined in economic quantities. This was because aluminium, one of the most abundant metals on Earth and is a very chemically-reactive element, strongly bonding to oxygen as the mineral bauxite, a complex oxide-hydroxide. It is a very difficult metal to extract and was not isolated as a metal until 1825 in an impure form by Danish physicist and chemist **Hans Christian Ørsted** (1777-1851).

Bauxite is formed by the complex weathering of feldspars, often mixed with a variety of iron oxides and hydroxides, in tropical regions with distinct wet and dry seasons which produce a red-yellow lateritic soil. This is usually strip mined and the naturally-rounded pellets of bauxite are then sent for processing and then smelting.

Processing begins by converting the impure bauxite into pure, white aluminium oxide (alumina – Al_2O_3) using the Bayer Process. This involves several steps:

1. Digestion by washing with a hot solution of sodium hydroxide, (NaOH), at 175 °C, under pressure converting the impure aluminium oxide of the bauxite into soluble sodium aluminate:

$$Al_2O_3 \quad + \quad 2\ NaOH \quad \rightarrow \quad 2\ NaAlO_2 \quad + \quad H_2O$$

Al_2O_3	2 NaOH	2 NaAlO$_2$	H$_2$O
aluminium oxide	sodium hydroxide	sodium aluminate	water

This also dissolves any silica present and sometimes lime (calcium oxide) is added to assist this removal as calcium silicate.

2. Filtration is then used to remove the solid impurities, using a rotary sand trap, and a flocculent (which gathers fine particles together so that they can be trapped) such as starch. This waste mixture is called red mud.

3. Treating the alkaline sodium aluminate solution by seeding the supersaturated solution with crystals of pure aluminium hydroxide ($Al(OH)_3$):

$$2\ H_2O\ +\ NaAlO_2\ \rightarrow\ Al(OH)_3\ +\ NaOH$$

water   sodium aluminate   aluminium hydroxide   sodium hydroxide

The excess sodium hydroxide solution from the previous step is then recycled.

4. Calcined by heating to 980°C so that the aluminium hydroxide decomposes to aluminium oxide (alumina), giving off water vapour in the process:

$$2\ Al(OH)_3\ \rightarrow\ Al_2O_3\ +\ 3\ H_2O$$

aluminium hydroxide   aluminium oxide (alumina)   water

The white alumina powder can then be sent to the smelter. These are usually located near major electrical-producing centres near coal fields or hydroelectricity plants. This is because a large amount of electricity is required for smelting alumina to aluminium. For example, in Australia, most of the bauxite is mined at Weipa in the far tropical north of the country, but the smelters are far to the south near sources of coal-fired power stations at Gladstone, Queensland; Newcastle, New South Wales; and Portland, Victoria; or in Tasmania which has hydroelectricity in the central mountains and a smelter at Bell Bay.

The most common method of smelting alumina is the Hall-Héroult process, which involves the dissolving of the alumina in molten cryolite (sodium aluminium fluoride - Na_3AlF_6) and electro-refining the molten solution in a purpose-built **electrolysis** cell.

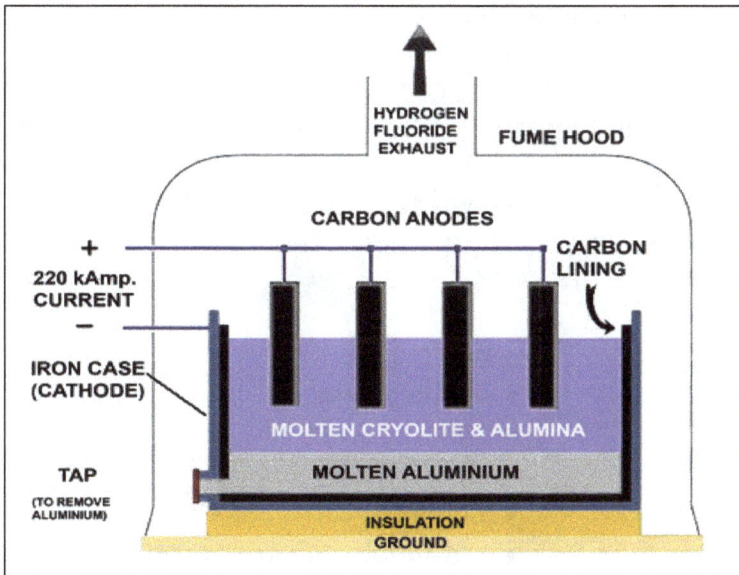

Figure 5.19: A simplified diagram of the Hall-Héroult electro-refining process

Chapter 6:   Fuels and Energy

6.1 Introduction

In an industrial world, materials from the Earth have become more important in providing the raw materials for the production of energy. This also comes with a cost to the environment, both in damage to the land surface, pollution of natural waters and the air, and contributing to global warming.

The major fuels which are used today on an industrial basis are non-renewable and are:

- fossil fuels such as coal, oil and gas
- nuclear fuel derived from the use of uranium metal

Renewable fuels which have less impact on the environment and provide the security of future availability include:

- hydro-electricity
- solar power
- wind power
- tidal power
- biogas
- ethanol and other biofuels
- geothermal
- hydrogen fuel

6.2 Fossil Fuels - Coal

Coal is formed from the **anaerobic**, or without air, breakdown of large volumes of plant matter which have fallen into the quiet waters of freshwater swamps or **paludal** environments and lakes or **lacustrine** environments.

Coalification is the process which turns compressed vegetation into coals with the loss of water (H_2O), carbon dioxide (CO_2) and methane (CH_4). The extent and rate of this process depends upon:

- type of original vegetation

- depths of burial

- temperature and pressure at these depths

- length of time

With time and the temperature and pressure associated with burial, compressed vegetation begins to transform into various **ranks** of coal. For purposes of coal combustion, ranks are described in terms of their components from a coal assay which is a laboratory analysis its components. The main components of coal are:

- Fixed carbon which is the useable organic fuel component as carbon.

- Ash which is the silica content which will be left over when the coal is burnt and which also contains some toxic materials such as the salts of arsenic, lead and mercury.

- Moisture which is the remains of the water expelled by the coalification process which hinders burning but is liberated as steam during burning.

- Volatiles, except for moisture, are usually a mixture of short and long chain hydrocarbons, aromatic hydrocarbons and some sulfur.

Generally, the rank of coal increases with depth if the thermal gradient or increase in temperature with depth, is entirely vertical. This relationship is called **Hilt's Law.** However, there may be some changes in rank due to other external changes, such as the heating effects of local igneous intrusion or pressure by large scale rock folding. The main ranks of coal are:

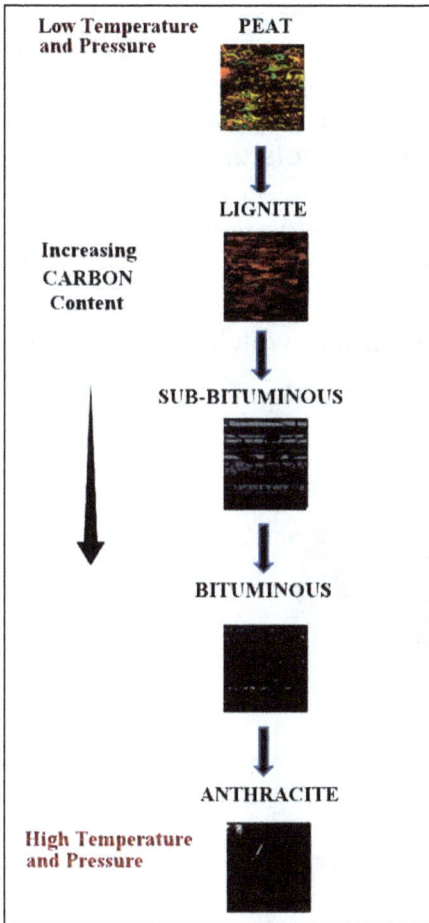

Figure 6.1 Ranks of coal

Peat is often considered only a precursor to coal as it consists of considerable vegetable matter with little coalification, high percentage of water, silica and clay (as soil). It is found extensively in parts of northern Europe in large areas of marsh and bog. It has, however, been a source of poor fuel for centuries and is still dug for local consumption in Ireland and Finland.

Lignite (or brown coal) is the lowest true rank of coal, having about 25-35% carbon but still high in moisture content and amount of silica (or ash). It is useful for generating heat in power stations

Sub-bituminous coal has 35-45% carbon content, about 15-30% moisture and considerable ash content. It is not suitable for making coke, the solid fuel made from distilling coal, but it is useful as a fuel in steam boilers. Together with lignite, this rank makes up the largest reserve of the world's coal.

Bituminous coal is the typical hard, black coal used in making coke for industry and most high-grade steam-generation boilers. It contains bright and dull bands and contains 45-86% carbon with some moisture and ash.

Anthracite is a very hard, often glossy in lustre, and may show conchoidal fracture as it is very brittle. It has 86-98% carbon and is considered to be a transition towards pure carbon as graphite which forms after further changes by heat and pressure.

Figure 6.2: Lignite or brown coal showing woody texture (right)

Figure 6.3: A banded bituminous coal showing bright bands called vitrain

Figure 6.4: Anthracite coal showing conchoidal fracture

About three metres of compressed vegetation will eventually form one metre of black coal, and because of the nature of the deposition of the vegetation into freshwater lakes and swamps, the coal is formed in layers called seams. These are found within the originally horizontal sequence of sedimentary rocks, usually with shales with sandstones and conglomerates as coal measures. Seams can be many metres thick and the area of the seam may be several hundred square kilometres. Coal measures may contain several seams, representing different periods of deposition. Much of the world's coal was formed in the Carboniferous Period (359.2 to 299 million years ago), although considerable coal formation occurred in more recent times in the Permian Period (299 to 251 million years ago) and the Triassic Periods (251 million and 199 million years ago), especially in the southern hemisphere.

Figure 6.5: Sedimentary strata showing a sequence of shales and sandstones at top and a coal seam at base from the Late Triassic Ipswich Coal Measures, Brisbane, Australia.

Non-banded coals are clean, compact blocks of massive structure with fine-grained texture. Usually they are dark grey to black, have a greasy lustre, and a marked conchoidal fracture. They contain none of the bright and dull bands seen in banded coals, but are black and granular and consist of large quantities of plant spore outer layers, pollen, and algae. If the algal content is low, the coal is called **cannel coal**, but if it is high, it is a called **boghead coal**.

Coal can be mined by several methods, and coal mining has been a major industry since the 18th Century in Europe and America. The main techniques of coal mining are:

- Strip mining by mining on the surface as huge open-cut strips if the coal seams are close to the surface, are mainly horizontal, and cover a sizeable area.

- Bord-and-pillar method which is also called the room-and-pillar method. This is a sub-surface method used when the seams are deep, close to centres of habitation or greatly tilted and faulted.

With strip mining, the area mined is often extensive in area but shallow in depth and such an operation requires a large capital investment initially, but results in high productivity, lower operating cost and good safety conditions. Because of their size and relative ease of access, these mines use some of the biggest excavation equipment constructed, such as dragline excavators, which are constructed on site and are electrically-powered by on-board or nearby generators. For smaller-scale digging, bucket-wheel excavators are used. As with other surface mines, the vegetation and topsoil are removed first and stockpiled for later rehabilitation of the area. The overburden, the unwanted rock material which is removed, is usually back-filled as the excavation slowly moves across the mine lease. Eventually, when the mine is closed and the site filled, the land is restored.

Figure 6.6: Dragline (right) at an open pit coal mine, Queensland, Australia

Figure 6.7: Bucket-wheel excavator (Photo: CSIRO)

Online Video 11.1: Visit an open-pit coal mine in operation in Queensland, Australia.
Go to https://youtu.be/9Z9a02oGfBQ

Bord-and-pillar is a form of underground mining in which the coal is removed along tunnels or bords, by small electric excavators and skips leaving pillars to support the roof. Often roof support is also supplemented by roof bolts inserted with epoxy glue into holes drilled in the roof. As the end of the section being mined, the goaf area, the pillars are also mined, and the roof is allowed to fall in under control. Often the seam of coal is relatively thick, perhaps a metre or two, and horizontal. Then it can be excavated using the **longwall** technique in which a temporary track with hydraulic roof supports are laid along the face of the seam so that an automated excavation wheel can slowly move along the track, excavating the coal to a conveyor belt behind. As the first few metres of the seam have been removed horizontally, the whole system is then moved forward so that more coal can be taken out.

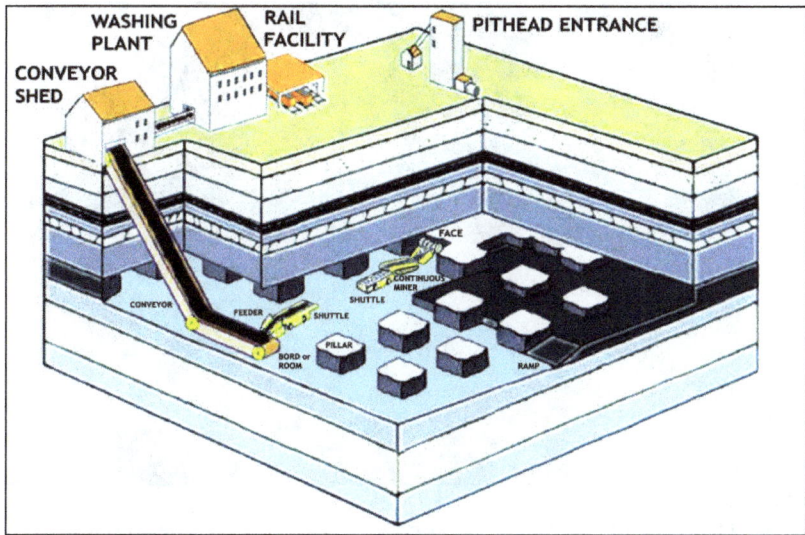

Figure 6.8: Diagram showing the bord-and-pillar method of subsurface coal mining

Figure 6.9: The pithead, or shaft entrance to a coal mine at Wollongong, New South Wales, Australia.

Figure 6.10: A Shuttle car used within the mine. Note the compact profile as the ceiling in the mine is less than two metres high.

Figure 6.11: Longwall excavator – notice the hydraulic ram supports at top right (Photo: Peabody Australia)

6.3 Fossil Fuels - Oil and Gas

Crude oil and the natural gas which often comes with it are formed from the vast quantities of microscopic dead marine organisms which fall into and accumulate in the muds or ooze, of the deep oceans. In time, this is covered by more marine sediments, mainly muds and sand, and the organic debris goes through complex changes under anaerobic conditions in the absence of air to become oil. If the ocean floor is uplifted to become land or shallow sea basins, the oil, along with gas and trapped water will migrate upwards from its source rock until it reaches the surface, where it might form tar pits, or become trapped by an impermeable cap rock in a variety of geological structures collectively known as oil traps.

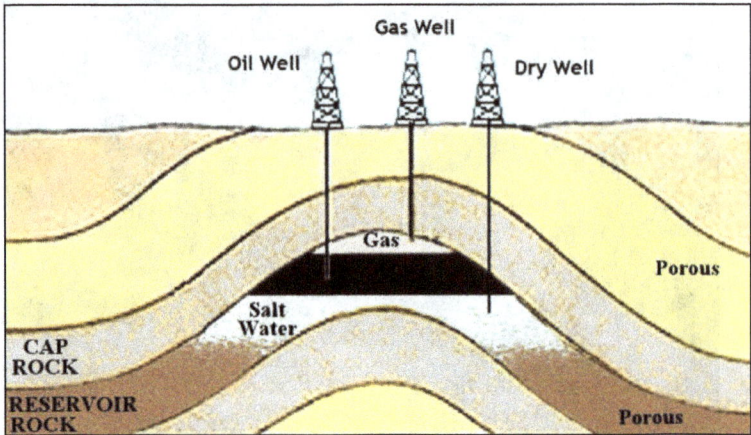

Figure 6.12: An anticlinal oil trap – more accurately it is a 3-Dimensional dome

Other oil traps include:

- Fault traps in which a fault, caused by movement along a joint or crack, has dislocated strata such that impermeable (water and gas proof) layers slide up against permeable reservoir rocks, trapping the fluids.

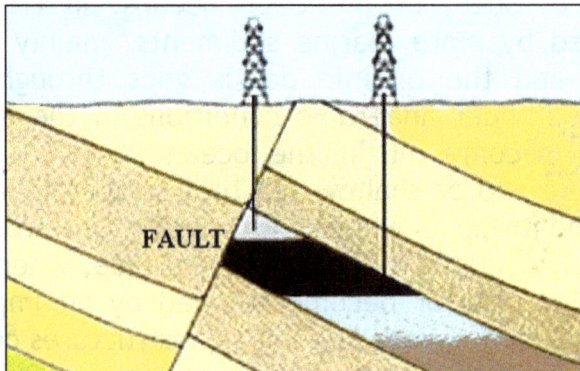

Figure 6.13: A fault oil trap

- Stratigraphic traps in which a cap rock layer has been put down horizontally on top of a tilted and confined reservoir rock making a structure called an **unconformity**.

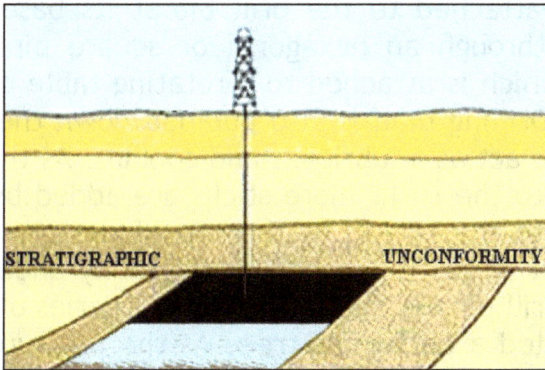

Figure 6.14: A stratigraphic oil trap

- Salt dome trap of impermeable rock salt may be squeezed up by tectonic forces forming a structure called a **diapir**, a name from the Greek: *diapeirein*, to pierce through- pushing up both reservoir and cap rocks.

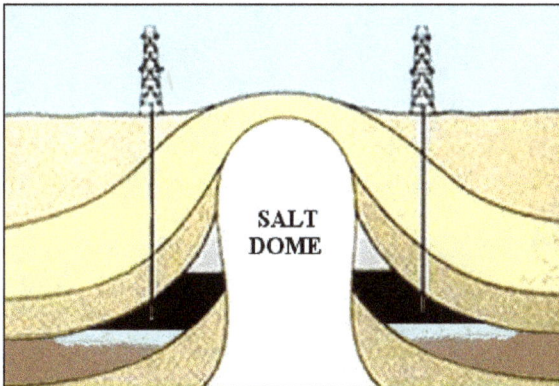

Figure 6.15: A salt dome oil trap

Oil rigs either on land, or on drilling platforms at sea, are essentially the same – a tall derrick up which the sections of pipes, called **sticks**, are hauled so that they can be attached to the drill bit at its base and then passed through an hexagonal or square pipe called a **kelly**, which is attached to a rotating table driven by a motor. Drilling mud is also pumped down the drill pipe casing to act as a lubricant and coolant. As the bit drills down into the rock, more sticks are added by screwing them into the thread at the top of the previous pipe. When oil or gas is reached, it is usually under pressure so the drill pipe is then attached to a series of pipes and taps called a **christmas tree**. As the pressure reduces, the oil or gas, stops flowing under its own pressure and must then be pumped out by injecting water down into the well.

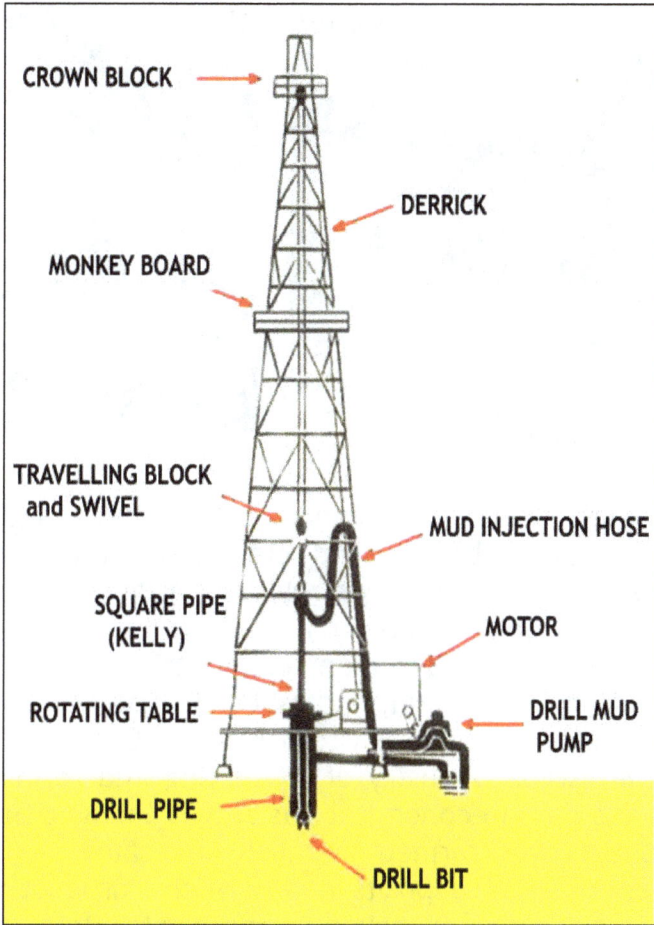

Figure 6.16: Diagram of an oil drilling rig showing its main parts

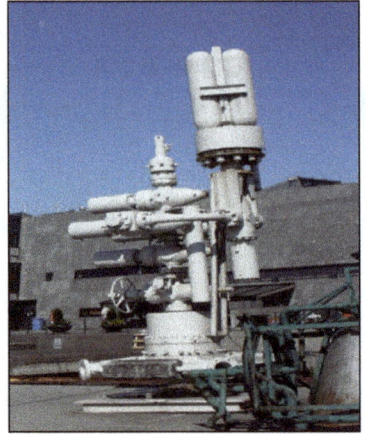

Figures 6.17 & 6.18: a drill bit (left) and a christmas tree (right) outside the oil museum, Stavanger, Norway

Oil sands are loose sands or broken sandstones containing a mixture of sand, clay, and water, saturated with a viscous form of petroleum (bitumen). It is also commonly and incorrectly called tar, and large quantities are found in many countries including Russia, the USA and Venezuela, and especially the Athabasca oil sands located in north-eastern Alberta, Canada. Canada has the largest industry of extracting and refining the oil sands on an economical basis. It is thought that this petroleum was formed in much the same way as in conventional oil deposits but here the product of the marine environment moved to another location and was trapped in coarse river sands. The lighter petroleum products evaporated away and the sand-clay-bitumen deposit was covered in new sediment.

6.4 Fossil Fuels - Coal-Seam Gas (or Coal-bed Methane)

Coal-seam gas is found within the cracks and cavities of sub-surface coal seams. With the future threat of depletion of oil, this has become a new source of fuel which can be extracted from the surface without extensive mining. Coal seam gas contains about 95-97% methane gas (CH_4) and a little carbon dioxide, ethane and nitrogen. As a comparison, natural gas from oil wells also contains methane along with heavier hydrocarbons such as ethane C_2H_6, propane C_3H_8, butane C_4H_{10}, pentane C_5H_{12} and higher carbon molecules. It may also contain some hydrogen sulfide (H_2S) and carbon dioxide which must be expensively separated or removed. Methane gas has always been an explosion hazard in underground coal mines where it has been called coal mine methane CMC or firedamp, from the German: *dampf* for vapour). It is released during the mining process and often extracted by large ventilation systems to limit the hazard of mine explosions. Methane gas is a product of the breakdown of the organic matter within coal and is stored within the coal matrix as a liquid adhering to

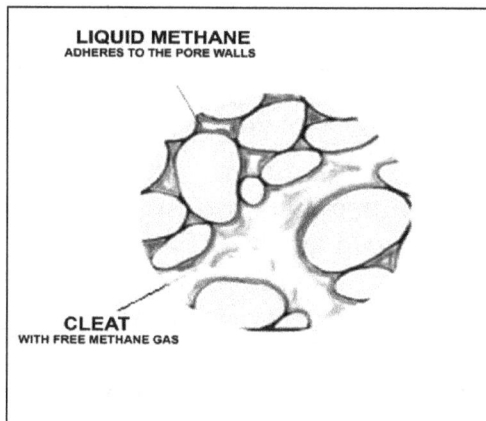

Figure 11.19: Diagram showing methane within the pore spaces of rock

the walls of the pore spaces, a process called adsorption. The open fractures in the coal, called the cleats, can also contain free gas or can be saturated with water.

Potentially commercial coal seam gas (CSG), containing mostly methane gas, is usually found in seams saturated with water. The underground water pressure also holds the methane to the pore walls. The CSG is extracted by means of wells which are drilled down into, and along the coal seams of depths of up to 1000 metres. When water is pumped out of the coal seams the confining pressure is reduced, leading to the evaporation of the gas so that it can be collected. Gas compressors are then used to pump released gas to a central processing facility where it is dried and transported along a high pressure pipelines to shipping points.

Where the gas is tightly held within the coal seam, hydraulic fracturing (**fracking**) is used to crack up the coal seam and surrounding rock to release the gas. This process involves the high pressure injection of sand-water slurry and sometimes a range of chemicals, into the coal seam. This not only fractures the rock but also holds the fractures open, thus releasing the gas. Some of the chemicals used in the extraction are potentially toxic, such as **BTEX** which is a mixture of benzene, toluene, ethylbenzene and xylene with sodium hypochlorite, hydrochloric acid, cellulose, acetic acid and disinfectants. The water extracted from these wells includes large amounts of natural sub-surface water and some of the water and waste chemicals which were injected down the well. The water which is extracted is pumped into surface ponds for treatment. The CSG extracted is then dried and pumped by long pipelines to storage facilities at market points.

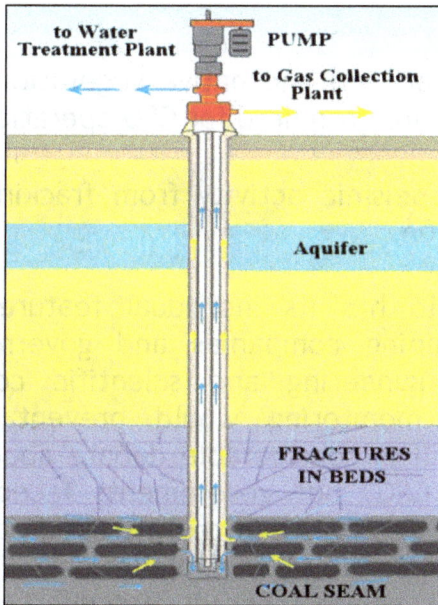

Figure 11.20: Coal seam gas well showing extraction and fracturing of beds due to fracking.

Environmentalists and farmers are concerned with the drilling of CSG in agricultural areas because of the potential damage to the environment and disruption to agriculture. They are concerned about:

- depletion and contamination of underground aquifers and surface water

- methods of disposal of the large quantities of extracted polluted water brought to the surface

- leakage of inflammable methane from wells, pipelines and storage facilities

- harm to humans and animals from air, water and soil pollution

- loss of agricultural land and native vegetation from the large surface area required for CSG operations

- risk of localised seismic activity from fracking and aquifer re-injection.

Each well and field has its individual features and properties, and mining companies and governments claim that the engineering and scientific controls through effective monitoring would prevent these problems. There still is considerable debate about the pros and cons of CSG, but governments faced with depletion of energy reserves and the attraction of CSG exports have welcomed CSG exploitation. Further development of environmentally-friendly energy sources, such as solar-electric power would be desirable not only to reduce the environmental impact but because CSG is also non-renewable.

6.5 Fossil Fuels - Advantages and Disadvantages

Humankind has long used fossil fuels as a source of energy. From the pre-industrial revolution to today, nations have used timber, coal, oil and gas to provide energy for heating, processing of raw materials and production of manufactured materials including synthetic materials directly produced from oil and coal, and fuels and oils for transportation. At a time when many are looking for alternatives, one must consider the advantages and disadvantages of these fuels.

The advantages of fossil fuels have always been:

- Readily available – humankind has relied on fossil fuels since the dawn of time, and there has been a very long and extensive development in gathering and using these resources. Moreover, the industrial revolution evolved to the present day using machinery and technology for coal and oil. Countries which are rich in fossil fuels have large infrastructures designed and made for the extraction of fossil fuels and their societies have also developed assuming the continued use of these fuels.

- Easier to find and to extract - Fossil fuels are relatively very easy to find using simple geological exploration. They occur all over the world, usually in very rich seams or oil traps and that once committed to getting coal or gas out of the ground or out from under the sea, nations have access to a considerable amount of energy resources.

- Extremely effective - Fossil fuels are extremely effective in producing energy with a known and generally reliable technology. This means that they can generate huge amounts of energy which cannot yet be matched by any renewable energy.

- Easier to transport – Fossil fuels such as coal and oil are relatively easy to transport in bulk by rail, road and by sea with minimum risk compared to the transportation of nuclear fuel. Renewable fuels cannot be transported although their energy can in the form of electricity. This can be a problem when considering long distance or from country to country, whereas fossil fuels can be transported to local power stations.

- Reliable - unlike renewable energy sources which have a dependency on Sun, wind and water, fossil fuels are readily available. A fossil fuel plant can be set up anywhere in the world using existing technology and human resources.

- Cheaper than most other sources – fossil fuels have a long-established technology and supply and are cheaper than most other energy sources. However, there are on-going costs of fuel supply and use, whereas hydroelectricity is more economic once the power plant has been constructed. This is also true of many of the other renewable sources but one has to consider the total amount of energy produced on a cost/unit basis. The problem with the argument that fossil fuels are cheaper fails when one has to consider the cost to the environment and what these fuels when the price of carbon-capture from them is added to production costs. When this is done, alternative fuels become more competitive.

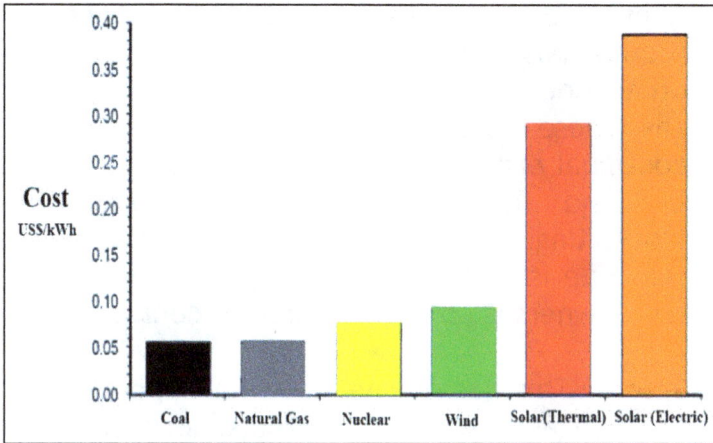

Figure 6.21: Chart showing some of the basic production energy costs of various sources without carbon-capture (Data: IEA/NEA)

- A major source of employment - Fossil fuels generate hundreds of thousands of jobs every year and many localities and economies rely directly or indirectly on the production of these fuels or their processing, transportation and use.

The disadvantages of fossil fuels are:

- Environmental pollution – this is the main disadvantage of using fossil fuels. The problems that are caused as a direct result of burning fossil fuels are well documented by scientists throughout the world. Carbon dioxide that is released into the air has been directly linked to global warming and its subsequent problems of sea-level rise, species extinction and social upheaval.

- Continued resources are needed - power stations require large amounts of coal and the transport and other industries require continuous supplies of oil and gas. These must be mined or extracted and so require continual exploration by geologists and large amounts of land use in their mining, processing and stockpiling. Land use for food supply as well as the sensitive issue of land rights by traditional and other land-owners is constant source of conflict.

- Safety- This has always been an important factor in the extraction of fossil fuels, especially in coal mining. This has been a dangerous occupation and more personnel are killed directly in coal mining and over time with lung diseases than in the nuclear industry.

There is considerable on-going research in the United States, China, India and Australia on various systems to overcome the problems associated with the use of fossil fuels. The burning of petroleum products in vehicles and that of coal in power stations are the major sources of greenhouse gases. Burning fossil fuels primarily produces carbon dioxide (CO_2) and water vapour (H_2O). Other major emissions include nitric oxide (NO) and nitrogen oxides (NO2) which together are called NO_x, sulfur oxides (SO_2), and soot (mostly inert carbon C). Much of the research focusses on eliminating these gases by more efficient combustion systems, their removal from the environment by burial or their conversion into new useful resources. Some of this research includes the following possible solutions:

- Redesigning of gas turbine and other combustion engines to work more efficiently with better fuels and combustion processes to produce lower gas emissions with increased power.

- Use of hybrid energy systems to produce electricity for general power and for transportation.

- Use of hydrogen gas as a fuel in both combustion engines and in **fuel cells** producing electricity for power in vehicles.

- Clean coal technologies using the vast amounts of coal currently being mined and in reserve and using much of the waste products as new resources. These systems may have the advantage of being able to be retro-fitted to existing power stations. Some areas of potential include:

- Removal of sulfur dioxide for conversion to useful sulfuric acid with **geosequestration** of carbon dioxide by pumping it underground into old saltwater aquifers. This uses current technology to dry the carbon dioxide and pump it to sites of old aquifers containing water too saline (salty) for other use. Here the gas is then pumped below the surface using existing drilling technology. Studies of the geosequestration of carbon dioxide from the Norwegian North Sea gas wells have shown this to be feasible. Since 1996, carbon dioxide gas has been pumped down 1000 metres into an old, depleted gas reservoir at the Sleipner well in the North Sea. Below

800 metres depth, the carbon dioxide liquefies and becomes trapped within the pore spaces. Moreover, interaction of the acidic carbon dioxide solutions in these reservoirs with local minerals, as cement or matrix in the pore spaces, would also help to seal in the carbon dioxide. There has been no reported leakage of any gas from this closely-monitored site.

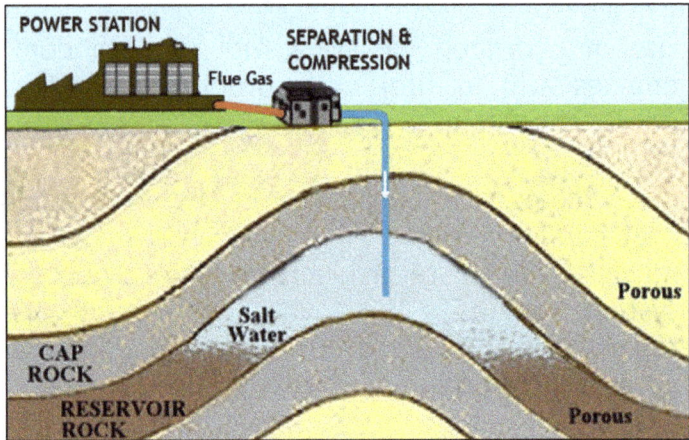

Figure 6.22: Geosequestration using saline aquifers

Sequestration into the deep oceans (say below 3000m) has also been suggested as a way of removing carbon dioxide gas from power stations. At such depths the gas liquefies and forms a pool which stays on the ocean floor as it is denser than sea water. This method may have some problems due to the possibility of disturbance of this pool by ocean-floor currents which may bring the concentrated carbon dioxide to the surface as gas.

- Exhausting flue gases from coal-fired power stations directly into ponds or photobioreactors containing algae which can use the carbon dioxide, even at high concentrations and elevated temperatures, to produce oxygen by photosynthesis. Flue gases typically contain about 82% nitrogen gas (inert), 12% carbon dioxide gas, 5% oxygen gas, 400 parts per million of sulfur dioxide gas, about 120 ppm of NOx and some soot. The latter can be precipitated as powder for land fill. Algae such as *Cyanidium celdanum, Scenedesmus sp., Chlorococcum littorale and Synechococcus elongates* can tolerate concentrations of carbon dioxide greater than 60%. There are some problems with this technology yet to be overcome before large-scale application can be made. These algae require large pond systems near the coast and there is considerable expense in land acquisition, construction and piping the gases from inland power stations. Closed systems using racks of pipes containing water with the algae seem to have more potential than pond or open systems which are also subject to massive water evaporation. In addition, the excess algae can then be processed into a hydrocarbon biofuel.

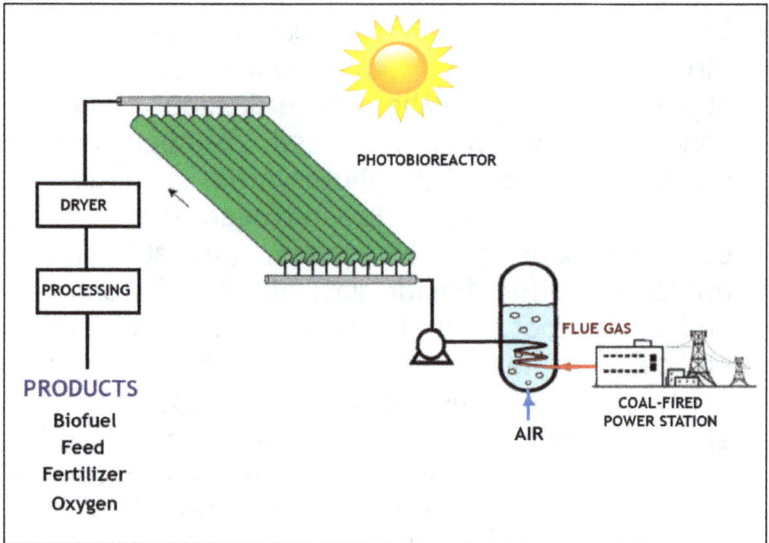

Figure 6.23: A simplified diagram of a photobioreactor used with flue gas from a coal-fired power station

- Application of old gasification technology to new systems which can produce many useable products and fuels with little or no gas emissions. In the early nineteenth century, coal gas (also called town gas) was produced by the direct distillation of coal in confined ovens. This produced an inflammable gas containing hydrogen, carbon monoxide, methane and volatile hydrocarbons together with small quantities of carbon dioxide and nitrogen. The process also produced coke, a soot-free solid used in home fireplaces and in making steel, tar, ammonium products and other organic chemicals. This gas was used in the home for lighting and cooking and in industry well into the twentieth century but was gradually replaced by electricity and natural gas.

Whilst some of the liquids and solids produced had limited use at that time, most were discarded into the environment as a toxic waste. Modern gasification systems are now under research and also currently coming into use in many countries. These operate with minimal or zero-waste gases so that any organic fuel added to the system will produce syngas (synthesis gas) for making synthetic natural gas (SNG), heat for steam for generating electricity and many other useful fuels and products. The raw materials for a gasification plant could include coal, oil or even some organic wastes. Useful products include sulfur, ammonia, hydrogen, methanol and other products already extracted from fossil fuels for the diverse industrial chemical industry.

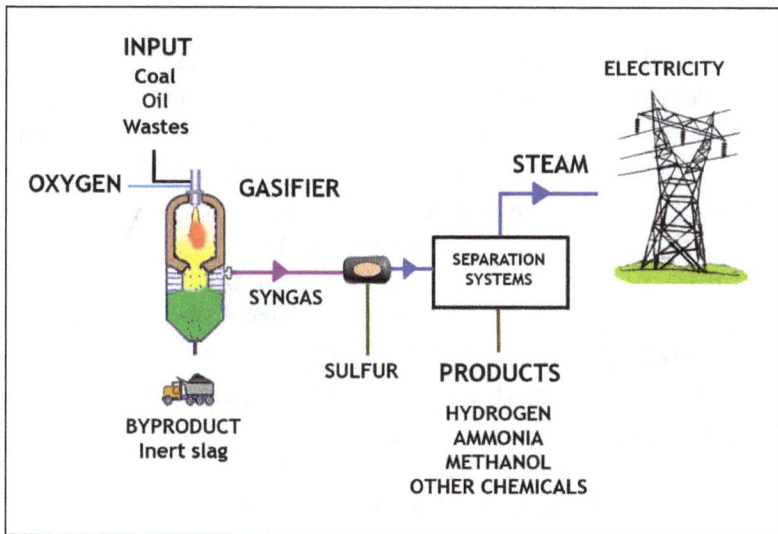

Figure 6.24: A simplified diagram of a gasification system

6.6 Nuclear Energy

This is the energy released by splitting the **nucleus** of atoms of uranium or plutonium by high speed neutrons in the process called **nuclear fission** within a nuclear reactor. This produces a chain reaction, as the **neutrons** from the nucleus of the atom are released by the fission of one atomic nucleus and then collide with other nuclei in other atoms and cause more fission, more neutrons, radioactivity and large amounts of heat. The heat produced by the reactor turns fresh water into steam which is then used to spin turbines coupled to electricity generators. The electricity is then regulated and passed along high tension lines to electrical sub-stations for conversion to the appropriate domestic supply.

The metal uranium, named after the planet Uranus, was discovered in 1789 by Martin Heinrich Klaproth (German: 1743-1813) in the mineral pitchblende (also called uraninite which is a mixture of mostly uranium oxide, UO_2 with UO_3 and oxides of lead, thorium, and rare earth elements). Uranium is the heaviest and last number of the 92 natural chemical elements. Its radioactive properties were discovered in 1896 by Henri Becquerel (French: 1852-1908), while working with phosphorescent material. The chemist Frederick Soddy (English: 1877-1956) first suggested in 1913, that the chemical elements each had different forms, called **isotopes**, which had slightly different weights whilst having the same **atomic number** and so the same chemical properties. This finding was based upon his studies of radioactive decay chains which showed that the unstable radioactive elements decayed through a series, to the stable element, lead. Uranium has eight isotopes, some artificially produced within reactors, but naturally occurring uranium is found as three major isotopes: uranium-238 (about 99% abundance); uranium-235 (0.72% abundance); and uranium-234 (0.0059% abundance). All three isotopes are radioactive, eventually decaying to lead with uranium-238 having a **half-life** of 4.4683×10^9 years. This half-life is the time taken for the radioactivity of the element to decay to half of its original level with the emission of radioactive particles and energy and the formation of other elements known as **daughter products**. Uranium-235 can be used in nuclear fission and has a half-life of 703.8 million years.

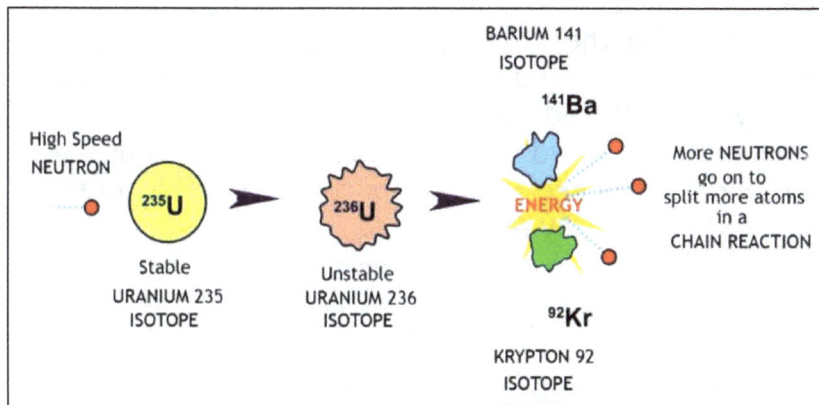

Figure 6.25: Diagram showing how uranium 235 is split to make energy. The numbers, such as 235, 236, 92 &141 are the atomic numbers of these isotopes

Uranium is mined in about twenty countries, with half of world production coming from just ten mines in six countries: in Kazakhstan (36% of total production); Canada (15%); Australia (12%); Niger (8%); Namibia (8%); and Russia (5%). The ores of uranium are usually mined by the open pit method because of the danger of radioactive radon gas which is often associated with the ores. After it is mined, the ore goes through a mill where it is first crushed. It is then ground in water to produce a mixture of fine ore particles suspended to form a slurry. This is then reacted with sulphuric acid to dissolve the uranium oxides, leaving the remaining rock and other minerals undissolved. These wastes are then run off into a mine tailings dam. This leeching process can also be done within the ore body itself by drilling and pumping the acid into the fractured ore with the solution then being pumped to the surface. The uranium slurry is filtered and the uranium extracted by precipitation, dried and then sealed in drums as bright yellow **yellowcake** (U_3O_8) concentrate. This uranium oxide is only mildly radioactive.

Figure 6.26: A specimen of uraninite ore (left) and the processed yellowcake (right) made from it

The refined yellowcake must still be further processed because the powder contains a mixture of uranium-235 and uranium-238, with only the uranium 235 being useful for nuclear fission. The yellowcake must then be enriched before it can be used in most power reactors. In this process, the proportion of the uranium-235 isotope is raised from the natural level of 0.7% to about 3.5% to 5%. This is done by series of reactions including the reaction of the yellowcake with nitric acid, ammonium hydroxide and hydrofluoric acid (HF) to form uranium tetrafluoride (UF_4) crystals. This is then treated with fluorine gas to form gaseous uranium hexafluoride (UF_6). This gas contains both UF_6 made from uranium-235 and UF_6 from the heavier uranium-238. The gas is put into a bank of **centrifuge** tubes which are then spun at incredibly high speeds, pulling the heavier uranium-238 UF_6 gas molecules into the centre of the tube for removal and potentially future conversion to plutonium for more fission. This leaves the lighter uranium-235 UF_6 gas molecules closer to the edges of the tube, where they can be extracted. The UF_6 gas extracted is then condensed to a relatively stable, white crystalline solid, which can later be vaporised in **autoclaves** (high temperature pressure chambers) and converted to the uranium oxide (UO_2) and excess gas being converted to more stable UF_4 crystals by adding hydrogen.

The idea that nuclear fission of the heavier, unstable elements could be used as an energy source was suggested in 1938 by the Germans Otto Hahn (1879-1968) and Fritz Strassman (1902-1980). Work on the application of nuclear fission developed throughout the early 1940s in the USA and in Germany and the first successful experiment in which nuclear fission first occurred was performed on 2nd December 1942, when the first **atomic pile** (Chicago Pile-1) began operations at the University of Chicago under the supervision of Enrico Fermi (Italian: 1901-1954).

This first atomic pile literally consisted of a pile or stack of large graphite blocks which acted as neutron moderators to slow down the neutrons and thus control the chain reaction. The fuel rods of uranium 235 oxide are inserted into the top of the reactor pile. When sufficient amounts of uranium in the fuel rods are put together, past its stable limit or its **critical mass** or about 52 kilograms of U-235, then a **chain reaction** occurs producing a great deal of energy, atomic radiation and daughter products of other, smaller radioactive elements such as barium ^{141} and krypton^{92}, are produced.

This experiment was part of the Manhattan Project undertaken by the US Government to develop an atomic bomb during the Second World War. In the 1950s, wartime application of nuclear fission turned to more peaceful uses, especially electrical power generation. Today (2016), there are about 440 commercial nuclear power reactors operating in 31 countries providing over 11% of the world's electricity supplies.

As well as energy and daughter products, nuclear radiation is also given off. The type of radiation depends upon the nuclear reaction and there are a great range of radiation types and conversions. There are three main types:

- **Alpha particles** (α) consist of a helium atomic nucleus of two neutrons and two protons. These have total electrical charge of +2 and an **atomic mass** of 4 units compared to the mass of one proton = 1. They have high ionization ability, that is, they can easily remove electrons from other elements to form ions, but low penetration power of only a few centimetres in air. They will not penetrate skin or clothing but are dangerous if they are inhaled or ingested with contaminated dust, food or water and cause tissue damage such as cancers.

- **Beta particles** (β⁻) are high-speed, negatively-charged electrons. Positrons (β⁺) which have the same mass as an electron but with a positive charge also occur in some reactions. Beta particles have very low mass, less ionization ability but moderate penetration of a few metres in air, to the base of the skin layer, and a few millimetres of aluminium metal. They will cause direct burning of the skin externally but more tissue damage can occur within the body including specific organ cancers if ingested with radioactive dust.

- **Gamma radiation** (γ) consists of high energy electromagnetic radiation similar to X-rays but more penetrating and dangerous. They have no mass but great penetrating ability, and will pass through the human body causing severe burns and great tissue damage. They are stopped only by several metres of concrete and about 20 centimetres of lead.

There are several types of nuclear fission reactors which are used to generate electricity coupled with steam turbines and electrical generators. A typical nuclear power station may have its nuclear pile contained within a steel pressure vessel within a larger concrete containment structure. Within the central reactor core are the fuel rods which contain the fuel. This is made from uranium dioxide (UO_2) powder by compacting it into cylindrical pellets which have been firstly heated to a high temperatures and ground to achieve a uniform cylindrical shape. Such fuel pellets are then filled into the metallic tubes made from either stainless steel or zirconium alloy. The zirconium rods are much preferred as they are corrosion-resistant and have low neutron absorption. The finished fuel rods are grouped into fuel assemblies that make up the core of a power reactor.

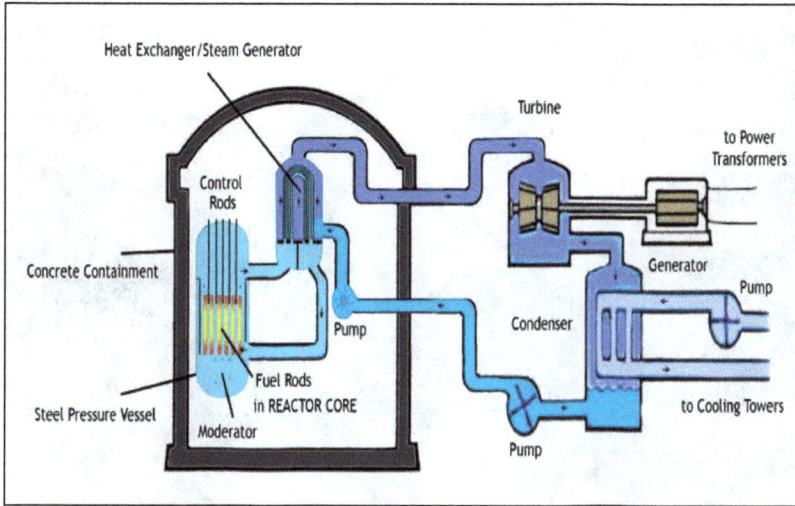

Figure 6.27: A diagram showing the main parts of a typical pressure vessel thermal reactor

In some pool-type reactors, **heavy water** is used as the **moderator** to slow down the neutrons of the chain reaction. This heavy water which consist of water molecules (H_2O) in which the hydrogen part is a natural isotope called **deuterium**, which has an extra neutron in its nucleus. These reactors use a large, deep, open pool filled with heavy water which acts as a radiation shield as well as the moderator. In the original atomic pile, graphite was used as the moderator but beryllium metal can also be used.

Figure 6.28: Looking into the core of a small, pool reactor used for research. The blue glow is due to Cerenkov radiation, which occurs due to electrons moving faster than the speed of light through the water moderator. (Photo: US Department of Energy)

Control rods can also be used to prevent the chain reaction happening too quickly. These can be made from an alloy of silver and cadmium or boron mixed with iron or carbon. These materials are good at absorbing neutrons and so reduce the number of neutrons available to continue the chain reaction. The control rods are moved up and down inside the reactor core to control the reaction. When lifted away from the fuel rods, they absorb fewer neutrons so the reactor gets hotter. If they are pushed down near the fuel rods, they absorb more neutrons and the reactor gets colder. By inserting them completely into the core, the control rods can used to shut the reactor down completely.

A small amount of nuclear fuel will make a large amount of energy. Energy is used over time, so usually the output of power stations is usually given in units of power which is the amount of energy used per time:

Power (in watts) = $\dfrac{\textbf{Energy}\ \text{(in joules)}}{\textbf{Time}\ \text{(in seconds)}}$

Nuclear power stations are not 100% efficient, so a lot of energy is lost to the environment and within the reaction itself. In general, a nuclear power station will produce about 1000 megawatts (MW) per day for only one kilogram of uranium fuel used. As a comparison, I gram of uranium will give about 1 MW per day, which is the energy equivalent of about 3 tons of coal or about 600 gallons of fuel oil per day which produces approximately 0.25 metric tonnes or 250 kg of carbon dioxide.

On an historical note, the unit of energy, the joule (J), is named after James Prescott Joule (English: 1818-1889) who discovered the relationship between mechanical work and heat energy produced by it, and the watt was names after James Watt (Scottish: 1736-1819) who developed an improved version of the steam engine.

The advantages of nuclear energy include:

- Low pollution – nuclear power plants do not emit large amounts of greenhouse emissions such as carbon dioxide and methane associated with fossil fuels, although some plants use cooling towers to cool the core and so produce some heat pollution. There is usually little or no impact on local water supplies and no requirement for large areas of land storage.

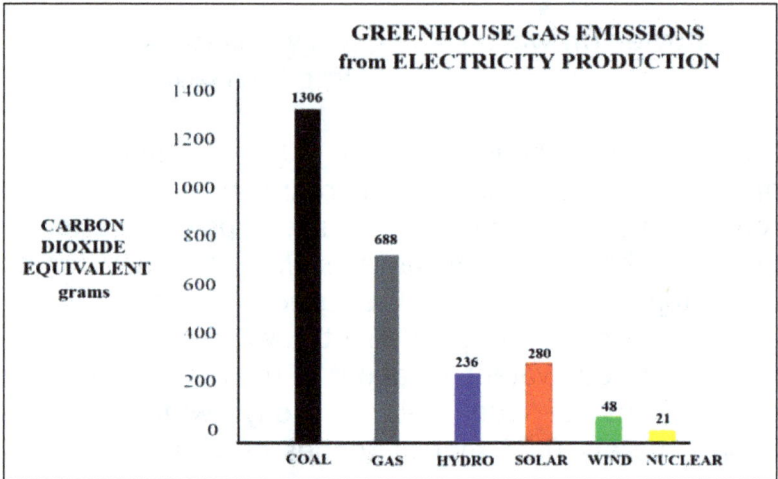

Figure 6.29 : Graph showing the relative levels of CO_2 emission (maximum) from generating one kilowatt-hour of electricity from different sources (Data: International Atomic Energy Agency 2000)

- Low operating costs - nuclear power produces very inexpensive electricity. The cost of building these plants is high, but once completed, the cost of fuel and continued operation is relatively low. The normal life of nuclear reactor is anywhere from 40-60 years, depending on how often it is used so they are very cost effective.

- Reliability – there are over 400 nuclear power plants operating in over 30 countries producing electricity for their national grid. The first plant opened at Calder Hall on the north-western coast of Great

Britain in 1956, and other countries soon followed. It has been estimated that at the current rate of consumption of uranium, the industry will have enough uranium for a reliable supply well into the future.

- More efficient than fossil fuels - the amount of fuel required by nuclear power plant is comparatively less than what is required by other power plants as energy released by nuclear fission is extremely large compared to that produced by the same amount of fossil fuel.

- Further potential - nuclear energy is not a renewable resource but, by using plutonium, a product of uranium fission in **breeder reactors**, more energy can be obtained from the amounts of waste nuclear fuel already available. Currently, breeder reactors are more expensive to build, require a higher critical mass, and often use liquid sodium metal, which is explosive in water and moist air, as a primary coolant. This makes such a reactor potentially more prone to accidents because of the dangerous coolant as well as radioactivity from core leakage. There is also a possibility of a safer source of nuclear energy in the future using **nuclear fusion.** This is the forced joining of the nuclei of the hydrogen isotopes deuterium and tritium to produce a helium nucleus and energy. Tritium is rare in nature but can be artificially produced in pool reactors, and has a nucleus consisting of one proton with two neutrons. At present research is attempting to find solutions for the containment of the very high temperature, electrically-charged **plasma** which would be produced in such a reactor.

The disadvantages of Nuclear Energy include:

- Environmental impact – this is perhaps the biggest issue in relation to nuclear energy and the mining, transportation, use and disposal of uranium and its daughter products. The process of mining and refining uranium has a number of issues, including the effects of radiation of the ore and associated radon gas at the mine site, as well as radiation dangers at all stages of its refinement. Transporting nuclear fuel to and from plants also represents a pollution hazard.

- Radioactive waste disposal - a nuclear power plant creates a considerable amount of nuclear waste per year. About 95% of the waste from electricity generation is high level waste (HLW), which gives off considerable amounts of dangerous radiation and high temperatures. Storage has always been considered a problem, and in the early days of nuclear energy, some was dumped at sea in sealed drums with potential for doing considerable damage to the ocean environment. Today, most waste is stored in drums in deep mines, especially stable salt mines, or in drill holes in stable rock areas. One area of potential for safer storage is synroc or synthetic rock, pioneered in 1978 by a team at the Australian National University. This involves the combining of the nuclear waste with titanium minerals and water which is then heated and fused into a rock-like substance. This would then be buried down very deep drill holes within stable crystalline rock and then capped with concrete, effectively returning the radioactive material to the Earth.

- High cost – whist still considered relatively economical to operate, there is still a considerable cost in the construction of nuclear power plants. Moreover, there is also a high cost associated with the eventual clean-up of the amount of radioactive waste kept in many current inadequate storage facilities. These administrative and environmental costs on top of the high expenses needed to build a plant, may also make it less desirable to use nuclear power in the future.

- Public opinion – there is considerable negative opinion in many countries which have nuclear power plants and in other countries which may consider the nuclear option. Environmental and health advocates are often active in preventing or limiting the governments to give the necessary permits for companies to set up new nuclear plants. The strong opposition to the nuclear industry in general, and the real or imagined threat of the use for breeder reactors to produce weapons-grade material, has become a major deterrent to many governments in the trading, use and waste storage of nuclear fuel. There is also some concern that unsecured nuclear material from any stage of the fuel processing and waste cycle may fall into the hands of terrorists.

- Potential for nuclear accidents – whilst the nuclear energy industry has an overall safety record superior to that of fossil fuels, several accidents have demonstrated that the effects are far greater in terms of long-term suffering, destruction of the environment and greater human impact. The threat of nuclear accident is always part of having power stations close to areas of human habitation. Accidents can be caused by human error and natural disasters, especially if the plants are located in geologically unstable areas such as in earthquake zones. Many countries rely upon nuclear power, and in countries with large populations but confined borders, the risk of a nuclear accident is always a major concern. To date, there have been three major peacetime nuclear disasters:

- Three Mile Island was a partial nuclear meltdown that occurred on March 28, 1979, in a reactor of Three Mile Island Nuclear Generating Station (TMI-2) in Dauphin County, Pennsylvania, United States. There was no loss in life as a result of this accident but it created alarm in the international public arena and cost a considerable amount in financial terms and confidence in the nuclear industry. The accident came about because of failure in valves operating the cooling system and the lack of sufficiently trained plant operators who were able to recognize the problem in time. The subsequent loss in coolant caused a partial meltdown of the reactor core as the heat suddenly increased. This also resulted in the release of radioactive gases, especially iodine into the immediate environment.

The repair and rehabilitation of the site did not finish until December 1993, with a total cost of about $1 billion. Luckily, the radiation which was given off did not seem to cause any significant cancers in the local population.

- Chernobyl was a catastrophic nuclear accident that occurred on 26 April 1986 at the Chernobyl Nuclear Power Plant in the city of Pripyat, which is now in the Ukraine. The event occurred because of a sudden and unexpected power surge which led to the rupture in one of the reactor vessels. This was followed by a series of steam explosions exposing the graphite moderator of the reactor into the air, causing it to ignite. The fire sent a cloud of highly radioactive dust, steam and gas into the atmosphere over an extensive geographical area including parts of Russia, Ukraine and Belarus. For the next four years, over 350,400 people were evacuated from the contaminated areas, especially from the heavily contaminated country of Belarus. The heroic struggle to contain the contamination and prevent a greater catastrophe ultimately involved over 500,000 workers and cost the lives of 31 people immediately with another 28 dying of radiation effects later. Further long-term effects, such as cancers and birth-defects are still being investigated and human suffering will be felt for some time.

- Fukushima Daiichi nuclear disaster was an accident at the Fukushima Nuclear Power Plant in Fukushima, Japan, caused initially by a **tsunami** which followed the Tōhoku earthquake on 11 March 2011. The reactors automatically shut down when the large ocean wave struck, but the tsunami destroyed the emergency generators which operated the cooling system, causing one of the reactors to overheat. This caused two other reactors to meltdown, the release of radioactive material, and several hydrogen-air chemical explosions. It was later found that the causes of the accident could have been prevented but that the operating company had failed to meet basic safety requirements and have adequate risk assessment and damage prevention plans, and proper evacuation procedures. Though there were no fatalities caused directly by the accident, it is expected that there will be about 130-640 people who will die of radiation-related cancers in the decades ahead. As yet, the decommissioning of the plant is still on-going and the plant management estimate that it may take 30 or 40 years to complete this task.

As a result of these three nuclear accidents, several countries have started a phase-out of their nuclear power plants. Sweden (1980) was the first country to start a phase-out, but this was repealed in 2009. Italy followed in 1987, Belgium in 1999, and Germany began the closure of 8 of its 17 reactors in 2000. Later Taiwan and Netherlands postponed their phase-out intentions. Austria, Denmark and Spain have enacted laws to cease construction on new nuclear power stations and the debate continues in Europe about the need for any nuclear phase-out. However, since closing down some of their reactors, some European countries such as Germany have re-opened or started construction of new coal-fired power stations to make up their energy loss and to meet new demands.

Many countries including Australia, Austria, Denmark, Greece, Ireland, Italy, Latvia, Liechtenstein, Luxembourg, Malaysia, Malta, New Zealand, Norway, Philippines, and Portugal have no nuclear power stations and remain opposed to nuclear power. Globally, more nuclear power reactors have closed than opened in recent years but overall capacity has increased.

6.7 Energy Production - Where to Now?

There has been considerable argument for and against the continued use of fossil fuels and the increased use of nuclear energy; especially now that it is apparent that the burning of fossil fuels has contributed to carbon dioxide emissions and that this has caused an overall global warming with a subsequent sea level rise.

Currently, the world faces a dilemma, with the demand for more energy conflicting with the problems caused by the burning of fossil fuels and the subsequent problems of global warming. Energy will be in great demand in the future, but there is a need to prevent any further rise in atmospheric and oceanic temperatures.

The table below (Table 6.1) shows that the most industrialised countries use the most power. The data in the last column (Average Power/Capita), can be misleading as many industrialised countries such as China and India also have a large rural population which uses much less energy than countries with a high city population. Norway, for example, ranking at number 27 has a per capita power consumption of 2603 watts/capita and only a population of about 5 million, however, over 90% of its electrical power generation comes from hydroelectricity and not fossil fuels.

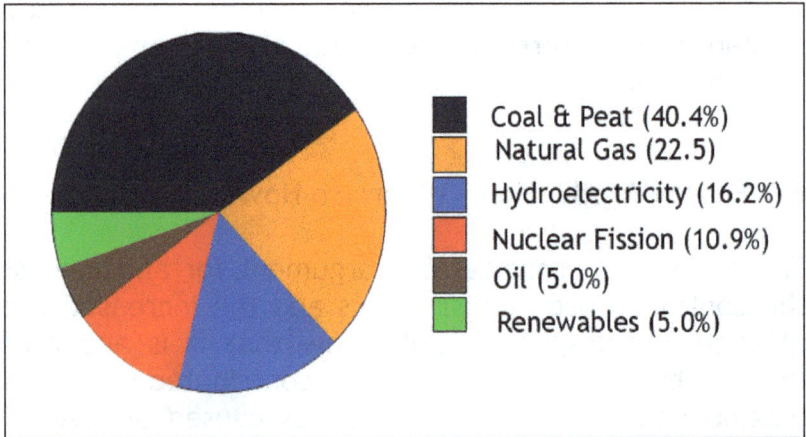

Figure 6.30: A pie graph showing the relative amounts of electrical energy produced by various fuel sources in 2012 (Data: International Agency http://www.iea.org)

Rank	Country	Electrical Consumption (MWattHour/ year)	Population	Average Power Use/Person (Watts/capita)
1	China	5,650,000,000	1,360,720,000	474
2	United States	4,986,400,000	317,848,000	1843
3	Russia	1,016,500,000	146,019,512	801
4	India	983,823,000	1,242,660,000	152
5	Japan	910,700,000	127,120,000	941
6	Germany	782,500,000	80,716,000	1160
7	Canada	570,800,000	36,584,962	2185
8	Brazil	493,500,000	201,032,714	301
9	South Korea	482,400,000	50,219,669	1038
10	France	451,100,000	65,864,000	904
11	United Kingdom	353,300,000	63,705,000	822
12	Italy	327,200,000	60,021,955	781
13	Turkey	264,136,780	78,741,053	335
14	Spain	253,100,000	46,609,700	875
15	Taiwan	224,300,000	23,315,822	1080
16	South Africa	234,200,000	54,002,000	415
17	Mexico	232,000,000	121,409,830	101
18	Saudi Arabia	231,600,000	29,195,895	781
19	Australia	229,600,000	23,060,903	1610
20	Iran	216,290,000	77,800,000	317

Table 6.1:   Showing the total power used by the top 20 countries.
(Data from "COUNTRY COMPARISON: ELECTRICITY - CONSUMPTION". CIA. 2013)

Some of the alternatives to fossil fuels and nuclear energy include:

- Hydroelectricity – is not really a new, alternative energy as it has been used since pre-industrial revolution days to power mills to grind grain and perform other, simple tasks. Similarly windmills have been used to perform relatively simple, mechanical tasks. Since the 1960's there has been a conscious effort to expand the use of water power to produce electricity which would not add to the pollution of the environment. Hydroelectric power generation is often the major source of energy in countries which have high mountains and large amounts of precipitation as rain or snow. It requires a large capital outlay in the construction of dams, pipelines and power stations, but once operating, costs are minimal and there is little damage to the environment. The water from these dams is piped downhill until it reaches sufficient speed to be passed into turbines within the power station. These are coupled to electrical generators which produce electricity.

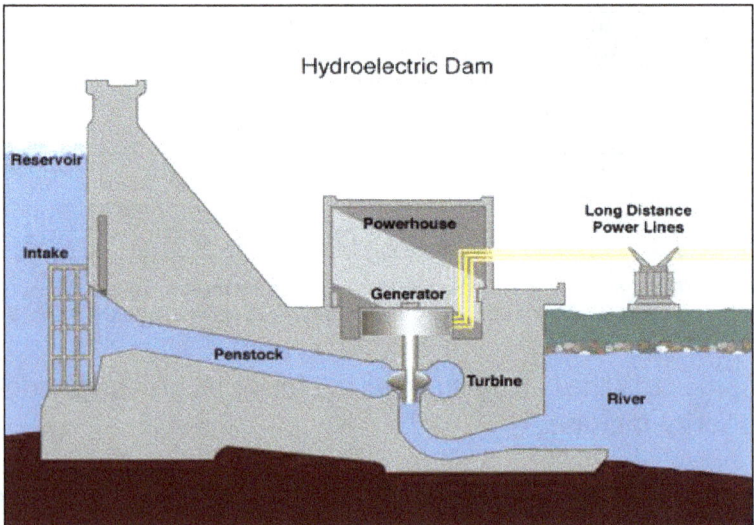

Figure 6.31: A diagram showing the operation of a hydropower station (Photo: Tennessee Valley Authority)

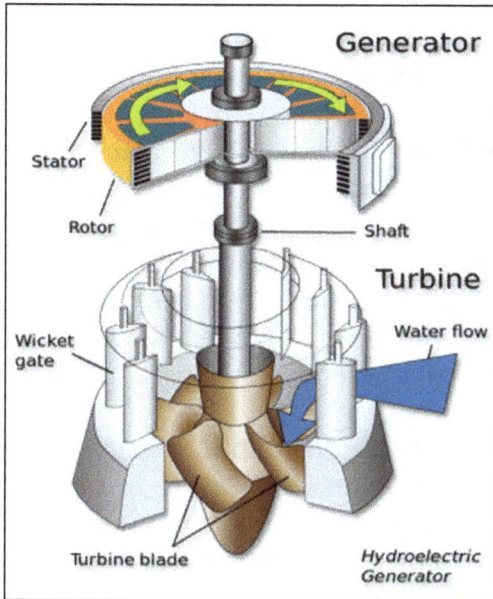

Figure 6.32: Diagram showing detail of the turbine/generator (US Army Corps of Engineers)

The water from the river which has been dammed usually continues on to be used further downstream. Only the lack of water and silting of the lake behind the dam are major problems to the generation of power but concerns have been raised in some countries about the restricted water flows downstream, the flooding of inhabited farming areas and the restriction of fish migration upstream. There is also a problem that some major lakes produced by dams may produce methane gas from organic sediment on the lake floor.

The Snowy Mountain Scheme in Australia, completed in 1974, and the Three Gorges Dam in China, operating since 2012 are just two of the world's major hydroelectric systems. In 2017, the Australian Government announced that the Snowy Mountain system is to be updated so that its 4000 megawatt system will be increased by 50% using a **Pumped Hydro Energy Storage (PHES)** system. This system makes better use of hydroelectricity production by pumping water back into the dams when electricity grid loads are not at peak requirement and then releasing the water for power production when more electricity is required.

According to the Renewable Energy Policy Network, in 2015, hydropower generated 16.6% of the world's total electricity and 70% of all renewable electricity, with an expected increase of about 3.1% each year for the next 25 years.

- Wind power like hydropower, has been used for centuries to drive small-scale industry such as grinding mills and to pump water. In more recent times it has been used on a larger scale to produce electricity, and most modern wind turbines can produce power with rated outputs of 1.5-3 MW. There are two main types of wind turbines: the classical Horizontal Axis Wind Turbines (HAWT) and the Vertical Axis Wind Turbines (VAWT). The latter type include the Savonius and Darrieus models, which have the advantage over the propeller HAWT wind turbine, in that they can operate in low winds coming from any direction. Propeller models need to be swung around into the wind to achieve maximum generation.

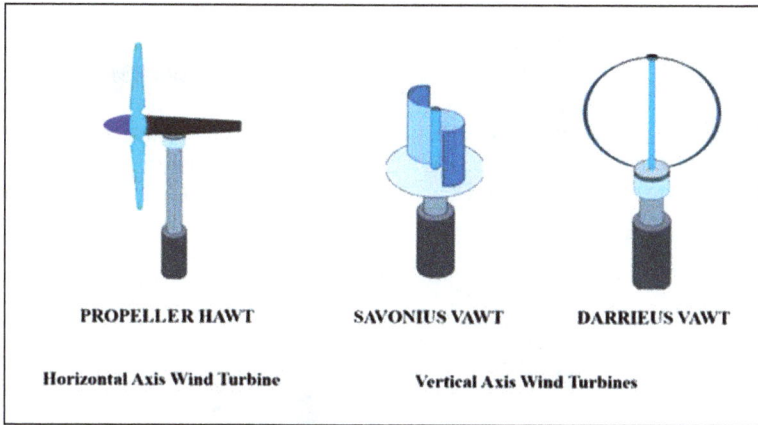

PROPELLER HAWT   SAVONIUS VAWT   DARRIEUS VAWT

Horizontal Axis Wind Turbine   Vertical Axis Wind Turbines

Figure 6.33: Diagram showing the most common wind turbines

There is still considerable potential for wind power and this would require a large-scale commitment to install towers over large areas such as wind farms in high wind environments such as offshore or wind-swept plains. Even then, wind turbines only have typical operating efficiencies of from 16 to 57% annually. In 2014 global wind generation was 706 terawatt-hours or 3% of the world's total electricity (BP Statistical Review – a terawatt is 10^{12} watts).

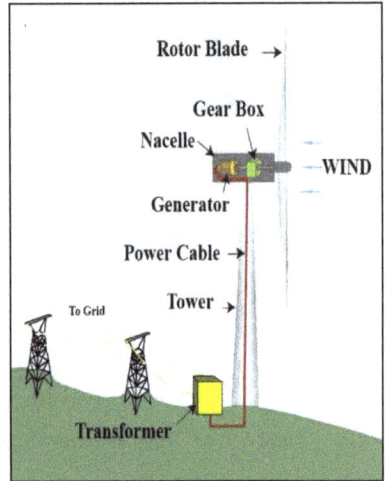

Figures 6.34 and 6.35: A propeller wind turbine in suburban England (left) and a diagram showing how such a turbine generates electricity

- Solar energy – has perhaps the greatest potential for future energy needs, especially in countries which receive considerable amounts of sunshine. Solar technologies are broadly grouped as either passive solar or active solar technologies, depending on how they use the radiant heat and light from the Sun. Passive solar techniques attempt to provide the maximum heating or cooling effect for existing conditions, such as orienting a building in respect to the direction of the Sun or building glasshouses in temperate climates to grow tropical plants. This could also mean selecting materials with favourable properties and designing the spaces within a building so that the heat of the Sun will allow a better current of air throughout. Active solar technologies are those which are constructed specifically to capture solar energy for direct heating or by way of conversion, to another form of energy such as electricity. Active solar technologies include:

- **Solar Thermal Energy** (STE) which uses solar collectors for heating water. Usually this is used for domestic or industrial hot water systems which heat water in metal or plastic tubes within a black, glass-enclosed box which then stores the hot water in an insulated water tank which retains sufficient heat over night. Usually the water tank is placed above the collector so that the water can rise by convection when it is heated and the whole system is placed high up, usually on a roof, so that the hot water can run down by gravity. If a mains water pressure is not available to replace water in the collector, then cold water must be pumped up daily. This is a simple operating system and home-made versions are easy to make, consisting of black plastic agriculture pipe coiled within a black-painted metal box covered with a sheet of glass or clear plastic and connected to a small water tank. Some domestic systems in colder climates which have limited daily sunlight also have an electric heating element in the water tank to boost the temperature at night.

Figure 6.36: A small domestic solar water heater

A similar system can be used in larger water heating applications such as the heating of swimming pools. In these systems, mats of heat-absorbent black plastic are used to cover a large area facing the Sun. Heated water is then re-circulated by a pump to warm the pool sufficiently to allow swimming in colder seasons.

Figure 6.37: Solar pipes on a rooftop for heating a swimming pool.

- **Concentrated Solar Power (CSP)** is an industrial application of solar heating which generates a very large amount of solar radiation by using mirrors or lenses to concentrate a large area of sunlight onto a small area. Here, the extreme heat may be used for industrial purposes such as the melting and refining materials, high-temperature chemical research, or producing steam for turbines to power electrical generators. CSP has being widely developed in Spain, Morocco, the Middle East and in the United States. The global operational power stands at about 5000 megawatts (MW) annually (see Table 5 for a comparison with total energy requirements).

Figure 6.38: A Concentrated Solar Power system used for research (Photo: US Department of Energy)

- **Solar electricity** is the production of domestic and industrial electricity using the rays of the Sun. The idea of using light from the Sun to produce direct current (DC). DC is a flow of electrons from the negative to positive electrical terminals developed from the discovery of the **photovoltaic effect** (PV) in 1839 by A.E. Becquerel (French: 1820-1891), the father of Henri Becquerel (French: 1852-1908), the discoverer of radioactivity. The Photovoltaic effect produces a voltage, or potential difference, which is the electrical pressure which pushes electrons through conductors producing a current.

In 1905, Albert Einstein (German: 1979-1955) described the photo-electric effect which is related to the photovoltaic effect but slightly different, as being the removal of electrons off a metal in a vacuum when it is struck by ultra-violet light. Einstein suggested that the **photoelectric effect** demonstrated that light was composed of small bundles of energy, called **quanta**. These acted like particles, now called **photons**, rather than continuous waves, which would knock electrons out of metals if the photons had sufficient energy. This discovery led to the concept of quantum mechanics now accepted in physics and earned Einstein the Nobel Prize in Physics in 1921.

The first solar cell using the photovoltaic effect which was able to produce a very small amount of electricity was invented by the Bell Laboratories in the United States in 1954. In simple terms, solar cells are made from a thin wafer of semiconductor, a material such as silicon which conducts electricity only at certain conditions, and which has been specially treated so that it becomes positively-charged on one side and negatively-charged on the other. When light energy strikes the solar cell, electrons are ejected from the atoms in the semiconductor material and form an electrical current when the two sides are connected by a conductor. The current produced is very small, only a few thousandths of an ampere (i.e. a few milliamps), so many cells are linked together on a flat surface called a photovoltaic module in series such that a negative side of one is attached to the positive side of the next and so on, so that the current of all the cells add together. The current produced by such a module is directly dependent on how much light strikes it and the length of daylight. Most domestic systems are only about 12% - 18% efficient.

Figure 11.39: A small 1.5 Kilowatt domestic photovoltaic module

For normal domestic electricity supply, the direct current must be turned into an alternating current, which provides energy by the to-fro motion of electrons, of the appropriate voltage using an electrical converter. This voltage varies from country to country with the USA using 110 V, Canada 120 V and many others using 220- 240 volts. The frequency of alternation also varies from country to country but it is usually about 50 to 60 hertz. The converters needed to change the direct, low current voltage from the solar panels to the main alternating domestic supply are a type of electrical step-up transformer. Whilst the converter increases the voltage, it loses some electrical energy in the conversion.

In many small applications, solar modules have been used to charge a wide range of items, ranging from pocket calculators, some domestic appliances and the International Space Station. However, the main problem with photovoltaic generation is that the amount of energy is relatively small compared to fossil fuel electrical generation, and that it only operates during days of good sunlight. Moreover, the manufacture of solar cells require a considerable amount of energy and material and if they are used away from the grid of the domestic power system, they require banks of heavy and expensive batteries to store electricity for night use. The efficiency of these batteries is improving but they are expensive, contain toxic chemicals which are difficult to recycle and may only last 3-4 years.

Figure 11.40: A large photovoltaic farm in the desert (Photo: US Dept. of Land Management)

Despite the problems of high cost, relatively low output, difficulty in storage of power and their modest use industry such the development of more efficient and cheaper solar photovoltaics is growing rapidly. Worldwide capacity reached 177 gigawatts (GW - 10^9 watts) by the end of 2014, which represents about 1% of worldwide electricity demand, with Germany producing over 7% of its national electricity demands (International Energy Agency). In addition, continued research into the storage of daylight electricity has shown some potential, especially in the use of cheaper, more efficient batteries and the production of hydrogen fuel by electrolysis. This is the splitting up of the water molecule by electricity into oxygen and hydrogen gases which can then be used in fuel cells to produce electricity or in combustion engines for general power.

- **Tidal energy** is a form of hydropower which uses the movements of the Earth's tides to produce electricity. This technology which is still in its infancy, yet it has the potential for future electricity generation. Tides are more predictable and consistent than wind energy and solar power but as yet their use in electricity generation has suffered from relatively high cost and limited availability of sites. These sites must have a high tidal ranges or rapid flow velocities and this is a major limitation but there is some hope that improved technology will allow for a greater potential than previously expected.

Tidal mills have been used both in Europe and on the Atlantic coast of North America and the earliest occurrences date from the Middle Ages, or even from Roman times. In these mills, the incoming tide was contained in large storage ponds, and as the tide went out, it turned waterwheels that used the mechanical power it produced to mill grain. In modern times this idea has been used to generate electricity in a manner similar to that of hydroelectric dams, and the world's first large-scale tidal power plant was opened in 1966 on the estuary of the Rance River in France. It is a barrage system using a wall across the estuary and its 24 turbines reach a peak output of 240 megawatts (MW 10^6 watts).

Figure 6.41: The barrage of the Rance River, Brittany, France (Photo: Wiki Commons)

Currently, the Sihwa Lake Tidal Power Station in South Korea, completed in 2011, is the world's largest tidal power installation, with a total power output capacity of 254 MW. It is near Inchon, South Korea and consists of a large barrage across a coastal inlet. A variation on this is to build a barrage across larger ocean bays, creating an artificial lagoon as a tidal lagoon system. Currently one such system is under construction within Swansea Bay in Wales. The larger size of these lagoon systems means that they can be built as multiple barrages and thus make use of the different tidal heights at different times of the day, spreading out their operating times.

The **tidal stream generators (TSG)** installed in these and similar projects, make use of the fast flow moving water to directly power turbines, in a similar way to wind turbines and using similar rotor types. These include the Propeller, Savonius and Darrieus blade types which are submerged just below water level. Tidal generators may be built into the structures of existing bridges or can be entirely submersed, thus avoiding concerns over the visual impact on the natural landscape. Land constrictions such as straits or inlets can create the high velocities needed to power these generators but usually only during times when the tides are running into or out of the inlet. Turbines can be orientated as either horizontal or vertical, and may consist of open or ducted blades and they are usually anchored near the bottom of the stretch of water where the tidal velocities are greatest.

The first generator of this type was the Seagen generator, installed in Strangford Narrows in Northern Ireland, in 2008. This generator produced 1.2 MW for between 18 and 20 hours a day while the tides are forced in and out of Strangford Lough through the Narrows. It was, however, only considered as a pilot program and its power generation capacity was considered insufficient and so it is now being decommissioned. Many more tidal generators are planned for a variety of coastlines which have fast tidal races. These include the original site in Northern Ireland and near Darwin, Australia. Unfortunately this technology suffers from the limitations of the local marine habitat which is usually shallow, offshore and subject to collision with a variety of natural and man-made objects.

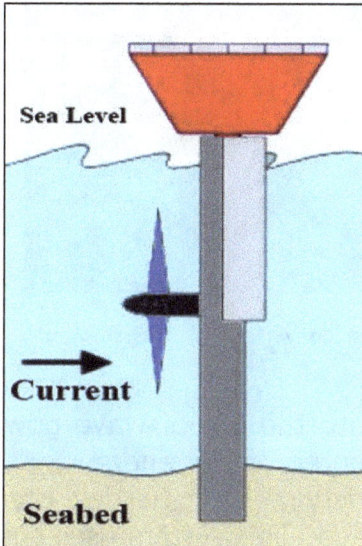

Figure 6.42: Diagram showing a Tidal Stream generator

- **Wave action generators** use the power of ocean waves created by winds and currents to generate electricity. A machine able to create wave power is known as a Wave Energy Converter (WEC). Wave-power generation is not currently widely employed commercially, although there have been attempts to use it since at least 1890. In 2008, the first experimental wave farm was opened in Portugal, at the Aguçadoura Wave Farm, and used three Pelamis-brand wave energy converters producing electricity of up to 2.25 MW. Unfortunately, due to some technical and problems with the parent company, the project was shut down only two months later.

Figure 6.43: One of the Wave Converters off the coast of Portugal (Photo: Wiki Commons)

Following further research into the use of wave power, the world's first grid-connected wave energy system began operations in 2016 offshore from Garden Island, near Perth in Western Australia. This system uses large, underwater buoys, or pods, which are anchored to the sea floor. The tube which connects these pods to their anchor also contains a vertical pump piston which moves up and down with the wave action. This piston is connected to an electrical generator which is capable of producing enough electricity to power over a thousand households.

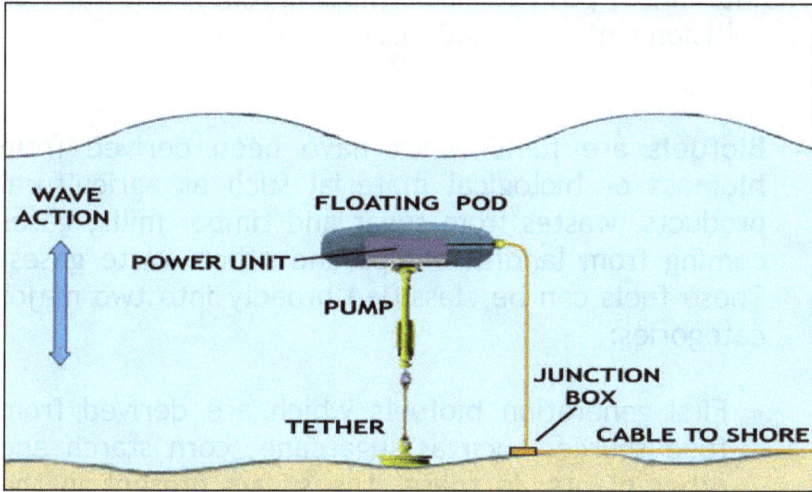

Figure 6.44: Diagram showing the operation of the wave action generator near Perth, Western Australia

The advantages of these wave action generators are that they:

- have little impact on the environment

- can be beneficial by acting as artificial reefs and encouraging marine growth

- can be brought on-line as required with predictable wave action

- can be used in a greater number of off-shore locations near coastal cities

The disadvantages of these systems concern their maintenance, position near shipping routes and the risk of collision with ships and marine mammals.

- **Biofuels** are fuels which have been derived from **biomass** or biological material such as agricultural products, wastes from sugar and timber mills, gases coming from landfill dumps and other waste gases. These fuels can be classified broadly into two major categories:

  - First-generation biofuels which are derived from food sources such as sugarcane, corn starch and other plants. In these, the sugars present in the biomass undergo **fermentation** and then are distilled to produce bioethanol, an alcohol which can be used directly mixed with petroleum in automobiles or used as a whole fuel in combustion engines or used directly in a direct ethanol fuel cell (DEFC) to produce electricity. Many countries which have a large sugar cane or sugar beet industry such Brazil, the United States, the European Community, China and Australia, use some ethanol in petroleum. However, the efficiency of ethanol as a fuel is only about 66% that of normal petroleum and there is still the problem of carbon dioxide air pollution, the competition for the source as a food product; and the expense and use of toxic metals needed in fuel cell construction.

- Second-generation biofuels use non-food-based biomass sources such as agriculture and municipal wastes. These biofuels mostly consist of **lignocellulosic** biomass, which is mostly non-edible, low-value plant waste consisting of lignin and cellulose fibres. Landfill waste dumps generate methane gas as a waste product of vegetable decay. Some waste dumps have been sealed so that the methane can be extracted and used directly as a fuel locally. It burns more cleanly than coal or oil, producing less carbon dioxide for the same amount of energy. Despite being a favoured alternative to petroleum, production of second-generation biofuel has not yet achieved the economies needed to fully replace petroleum and in some countries in recent years, production has actually dropped because of the competition with food and the availability of cheaper petroleum.

- **Geothermal energy** is generated in some volcanic regions such as the United States, the Philippines, Indonesia, New Zealand and Iceland where subterranean heat sources have been used to heat water, warm greenhouses and in home heating. More recently, such countries have used the high temperature steam from their geothermal springs to drive turbines for electrical power generation. Although the initial setup is expensive, and in some areas dangerous because of the proximity of volcanic steam near the surface, geothermal power is considered renewable and cost effective. Geothermal wells also release greenhouse gases trapped deep within the Earth, such as carbon dioxide, sulfur dioxide, hydrogen sulfide, methane and ammonia. These can react with precipitation to form acid rain in the local district and hydrogen sulfide gas and sulfur dioxide also have a very unpleasant smell.

Figure 6.45: The Nesjavellir Geothermal Power Station in Iceland (Photo: WikiCommons)

Figure 6.46: Pipes carry the high pressure steam to a conventional power station at the Wairakei Geothermal Power Station, north island, New Zealand.

On a smaller scale, there has been some limited success in non-volcanic regions which have deep, fractured crystalline rocks such as granite. These Enhanced Geothermal Systems use high-pressure water which is injected and pumped deep within crystalline rocks. At a great depth of up to five kilometres, these rocks are very hot because of the natural geothermal gradient of about 25°C/kilometre of depth. In some places, natural deposits of uranium ores in the crystalline rock also provide additional heat. The water which is injected into these wells, percolates through the fractured rock, is rapidly heated and then pumped to the surface. Here it passes through a heat exchanger which transfers this heat into a secondary system of pipes to heat fresh water and turn it to steam which can then generate electricity using a turbine coupled with a generator.

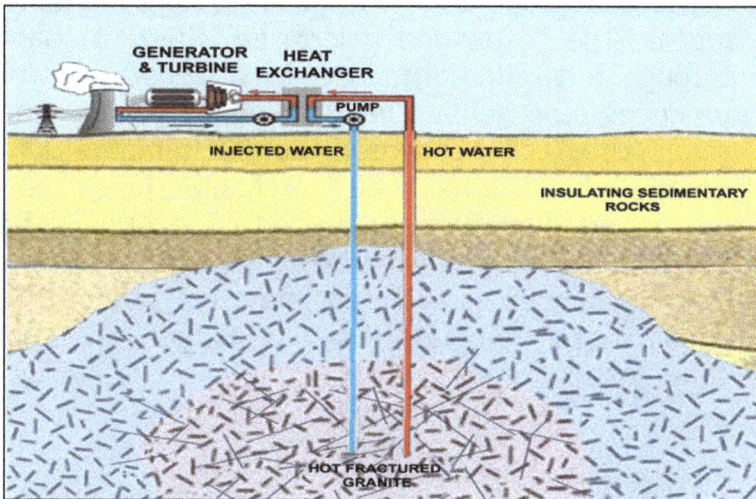

Figure 6.47: Diagram showing an Enhanced Geothermal System

Such systems can be used for generating electricity for local use, but they can be expensive and the water coming up from such depth is highly charged with mineral salts and can sometimes also be radioactive.

Another geothermal system which uses low temperatures (say 50°C to 350°C) typical of non-volcanic heat sources makes use of generators using an **Organic Rankine Cycle (ORC)** often referred to as the Binary ORC as it uses two systems of heat exchange. In simple terms, a fluid, usually water, is passed down pipes drilled in the geothermal reservoir which heats the water. On the surface, this heated water is passed through a heat exchanger where it heats and vaporises an organic fluid called the working fluid - often isobutane or pentafluoropropane - within a second system of pipes. The expanded vaporised fluid is passed through a turbine attached to a generator which produces electricity. The fluid which passes out of the turbine is condensed as its temperature is dropped by another heat exchanger and cooling tower and then recycled back to the original heat source. Such a system can be used with any relatively low heat source such as may be produced by decomposing biomass waste dumps and solar collectors.

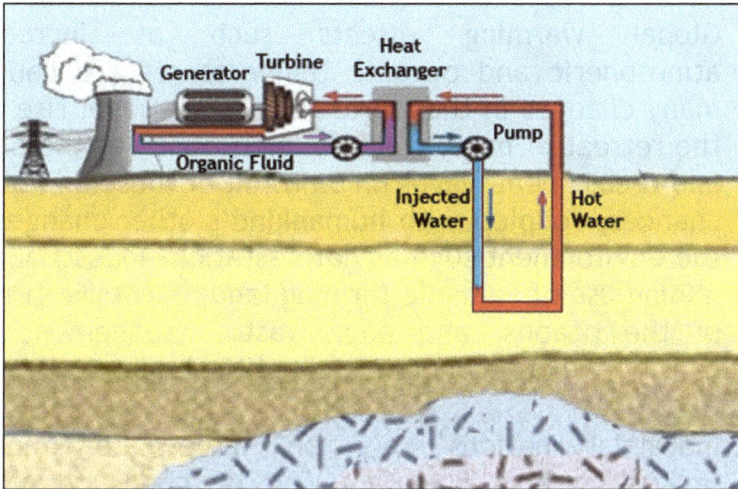

Figure 6.48 Simplified diagram of a Binary Organic Rankine Cycle power system.

6.8 Conclusion

Currently the global community is in a transition phase position between the old world of excessive energy use at any cost and a new world of cleaner, safer energy and a future which is less uncertain. In the past the main goal of production was linked to the need for increased material wealth and profits. Today the world faces considerable uncertainty but with some hope. To date, the situation appears to be:

- Global warming has come about because of excessive greenhouse gases, mainly carbon dioxide with methane, in the atmosphere, largely helped by the activities of humankind, especially since the 18th century at the beginning of the industrial revolution.

- Global warming effects such as increased atmospheric and oceanic temperatures are causing many changes in the world such as sea-level rise and the retreat of many glaciers, melting ice at the poles and record heat waves on land. All of these and other changes, coupled with humankind's other changes to the environment such as deforestation, industrial and mining use of valuable farming land, excessive fishing of the oceans, and poor waste management are causing social, economic and political upheavals.

- Industrial nations rely on large-scale mining, processing and manufacturing of goods for their economies. These activities require a large and continual exploitation of natural resources and the use of energy. Individuals who enjoy the benefits of these economies can practice less wasteful use of energy and recycling of resources and attempt to bring about social change. However, it is the large scale industrial and commercial applications at the national level which cause most of the global change.

- The world's industrial nations still depend upon the large-scale use of fossil fuels (coal, oil, and gas) and/or nuclear energy for their needs. Renewable energy sources are still at the lower end of total energy use.

- Most nations of the planet are now aware of the effects that humankind has had on the environment, especially the increase in global warming, and are aware of many of the issues of concern. It is hoped that, at the political level, these concerns will encourage reduction of pollution by fossil fuels and more research and development into the use of renewable materials and fuels.

In this transition period, more people need to be made aware of the actions of individuals and governments in causing these problems, and also being positive exponents of necessary change. Phasing out of all fossil fuels and industrial processes which put excessive greenhouse gases into the air will only come about in time. The pollution of the sea and land by mining, industry and agricultural chemicals will also require a great change in lifestyle and practice in the industrialised countries of the world.

Perhaps one scenario could be an intermediate period which may include the gradual reduction in emissions from the combustion of fossil fuels by the use of clean coal technology, natural gas and ethanol fuels, with further use of nuclear energy, possibly with the safe development of breeder reactors. This would gradually be phased out and replaced with large-scale use of renewable energies such as solar, wind and use of hydrogen as a fuel.

Whatever the future in the use of minerals, mined and processed materials and of energy resources, there will be a need for a corresponding shift in lifestyle, especially of those in an industrial world. Lifestyles may return to some of the simpler forms of transport, living with the environment rather than with artificial changes to it, and a better economy of resources.

Summary

1. Minerals are naturally-occurring chemical compounds having simple to complex structure and often a range of compositions.

2. The most useful mineral properties which are used in identification in the field are colour, habit, streak and hardness. A useful tool is a specimen of common vein quartz which is white and has a hardness of 7.

3. Crystal family identification can be difficult due to the variation of the exterior surfaces, twinning and the difficulty in finding crystals of a useful size.

4. The chemistry of minerals can be very complex due to the combining power (valency) of silicon (4) and ionic substitution.

5. The silicate group of minerals is the most common found in nature because of the abundance of the elements silicon and oxygen on Earth.

6. There are six members of the silicate group: nesosilicates (separate silicate tetrahedra e.g. olivines); sorosilicates (double-linked tetrahedra e.g. melilite); cyclosilicates (ringed tetrahedra e.g. beryls); inosilicates (single or double chains e.g. pyroxenes and amphiboles respectively); phyllosilicates (sheets of tetrahedra e.g. micas);

and tectosilicates (3D frame of linked tetrahedra e.g. feldspars and quartz).

7. Minerals are more easily classified using simple groupings such as chemical composition (e.g. silicates, carbonates etc.) which can be easily found using simple geochemical methods.

8. Economic minerals are those which are of value for use in industry and manufacturing. They are usually metalliferous ores, energy sources or used in building and manufacture of industrial and household products.

9. Apart from gold silver, sulfur and copper, economic minerals (such as ores) are usually chemical compounds as oxides, sulfides, carbonates, sulfates, phosphates and sometimes silicates.

10. To extract the wanted element (usually a metal) from its ore it must be first mined then processed, often by way of very complicated processes to extract the pure mineral from unwanted minerals (gangue) and rock. Then the mineral has to be converted and smelted.

11. Economic minerals are mined using many different processes, both on the surface and underground depending upon the type of ore, its location and proximity to markets.

12. Smelting is the final stage in processing the economic mineral before its marketing and use. Often smelting is a complex process involving further purification then the separation of the wanted element from one of its compounds (e.g. an oxide). This often uses considerable energy and chemical reactions.

13. A nation's wealth is often measured by its mining, mineral processing and production of the final product of these processes – gold, silver, iron, aluminium and uranium as well as by-products such as sulfuric acid and fertilizer which are important contributors to a nation's wealth.

14. Gems are any material of value that is used for adornment whereas gemstones are gems made from minerals and rocks. Both are valuable commodities for trading.

15. Fuel sources can be non-renewable, such as the fossil fuels (coal, oil, gas, nuclear fuel) or renewable, such as hydro, solar, wind, tidal, wave and biomass fuels.

16. Coal is formed by anaerobic (without air) compaction of large accumulations of plants under sediment from freshwater lakes, whereas oil and natural gas are formed by the complex reactions which occur within accumulations of dead marine organisms under the muds of the sea floor.

17. Coal mining and oil drilling is required to extract these resources from below the ground so that they can be further processed for use as fuels and raw materials for the manufacture of a wide range of vital materials including drugs, synthetic compounds, fibres and industrial chemicals.

18. Nuclear power depends upon the mining of uranium and its safe use within nuclear reactors to produce heat by nuclear fission (splitting of the nucleus of the uranium-235 atom). Generally safer than the fossil fuel industries, nuclear reactor accidents can occur because of human error and natural events and they are extremely catastrophic such as at Three Mile island (USA), Chernobyl (Ukraine) and Fukishima (Japan).

19. Fossil fuel combustion and their associated industrial processes, have largely contributed to global warming and harmful environmental change. However, alternative renewable energy and resources are yet to get to the levels demanded by an industrial world.

20. There is an urgent need to phase out all energy and manufacturing processes which use fossil fuel energy and replace them with environmentally safer energy sources and the use of renewable raw materials. This is slowly happening, but a major shift in social thinking is required if the World is to have a cleaner and more liable future.

Practical Tips

1. Fresh crystals are not easy to find in the field as they mostly weather very quickly. Quartz (as common white vein quartz and small hexagonal crystals in some granite areas) and orthoclase (as pink-brown blocky crystals in coarse-grained igneous rock) can be found as can gemstones, as small, glassy (often dull) pieces in river gravels carried from nearby eroded granites and similar igneous rocks.

2. Useful gemstones are noted for their hardness of 7 or more (e.g. topaz H=8, very clear transparent diaphaneity), so carry a pebble of common white quartz (H= 7) to test hardness.

3. Fresh surfaces (after some action with a hammer) will expose crystal faces in coarse igneous rocks e.g. small black rectangles of hornblende, stubby green crystals of augite, long, thin and shiny crystals of plagioclase.

4. Colour, hardness and lustre are the useful properties to know and test. Especially the metallic lustre of some ores, the hardness of 7 (greater in gems), and the distinct colours of copper (green), iron (red) and uranium (yellow).

5. It is handy to carry a small, secure bottle of sulfuric (battery) acid in a dropper bottle (CARE: highly corrosive). One drop on a clean surface can

often distinguish between white calcite (bubbles of CO_2) and quartz and pyrite/chalcopyrite (bubbles of smelly hydrogen sulfide gas) and gold.

6. The best ore and mineral specimens are often found in old, disused mines, but these areas are dangerous, having many disused shafts which may not be fenced and are usually unstable and subject to collapse. Caution is advised when walking in these areas. Also entering old mines is very hazardous due to poor timber supports and hidden shafts.

7. Mining and mine engineering and their many associated professions and trades pay very high salaries but often mining towns also have high costs of living. These professions are often very transient with mining booms and busts coming regularly - it is always a good idea to have a second, more stable profession in reserve.

8. Good localities for finding gems are in the outer bends of streams or in veins and vesicles (crystal-filled gas holes) in igneous or metamorphic rocks.

9. If interested in gems, one should join a local lapidary club which will offer assistance with prospecting, training in the skills needed to cut and set the gems, as well as opportunities to sell the finished product.

10. Economic minerals are often not easy to find because they tend to weather or are deep underground. A large part of the activities of geologists and geophysicists is devoted to the search for economic minerals as ores and fuel sources such as coal, oil and uranium.

11. Many of the important minerals as well as gold and silver are found in and around the edges of large igneous rock bodies such as granites. This is because the minerals often come up from deep below in these bodies or as hydrothermal veins in the surrounding country rock.

12. When these igneous bodies weather and erode, the valuable material is often carried away by local streams and forms alluvial deposits of such as gold, silver and diamonds. These can be won by panning and other density separation methods from beaches and rivers in places where the speed of the water is suddenly reduced, such as the inward bend of rivers (in coarse sands or gravels), at the base of rapids or waterfalls and in dune sands of beaches.

13. When panning for gold, silver or gems such as zircon always do so in strong sunlight to see the reflection of the item compared to the dullness of the bulk soil.

14. Some towns and cities are built on mining sites (this is why they were founded) and many old shafts and underground workings may still exist below new

home developments. Caution is required when considering building in these areas.

15. Because some economic minerals are poisonous and (e.g. lead and copper ores), weather to poisonous products and, or, are also associated with other poisonous minerals (e.g. of arsenic) one must be careful of the local water sources in current or old mining areas. In addition, many of the mine recovery processes used toxic chemicals (e.g. cyanide and mercury) to extract the metals and these may also contaminate the area.

16. The use of fossil fuels and uranium are the major sources of electricity. Reduction in their use can assist in combating further global warming. Domestic solar and wind energy devices are able to be used in the home and a simpler lifestyle with reduced use of personal automobiles will also assist.

17. Research into safer and renewable energy resources and the phasing out of fossil fuels and unsafe nuclear practices is to be encouraged.

Multichoice Questions

1. This question refers to the field sketch below, taken from a field geologist's notebook:

This mineral most likely has the property of:

    A. Conchoidal fracture
    B. Cleavage in one direction
    C. Cleavage in two directions
    D. Fibrous habit

2. An unknown mineral specimen will scratch fluorite but not orthoclase. The hardness of this unknown mineral is most likely:

    A. 3
    B. 5
    C. 6
    D. 7

3. A laboratory geologist, measuring the specific gravity of a mineral, suspends it from a mass balance and lowers it into water (density = 1.0 g/cm³). He/she obtains the following

$$mass = 18.8g$$

When placed on an ordinary balance in air, the specimen gives a result of 183.6g for its mass. From this data, and knowing that the apparent mass in water is equal to the volume displaced, he/she calculated the specific gravity of the specimen to be close to:

A. 8.8
B. 9.8
C. 18.8
D. 19.6

4. The following photographs are of common rock-forming minerals. Which of the following specimens has a perfect cleavage of three?

A.

B.

C.

D.

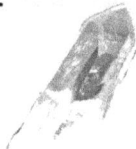

A. A
B. B
C. C
D. D

5. Sphalerite is an important ore of:

A. Tin
B. Nickel
C. Zinc
D. Magnesium

6. Minerals such as rutile (black) are concentrated on beaches and gold is concentrated in rivers. The property of both minerals which allows for this concentration is:

A. Metallic lustre
B. Crystal shape
C. Specific gravity
D. Cleavage

7. All significant banded iron formations (BIFs) were formed prior to 1800 million years ago. What does this indicate about conditions on Earth prior to 1800 MYA?

A. There was very little free oxygen in the
   atmosphere
B. Weathering of iron-rich rocks had not
   commenced
C. Deep ocean basins had not formed
D. Photosynthetic organisms were abundant

8. The geologist's sketches below show a section of hills
and the bedding within them:

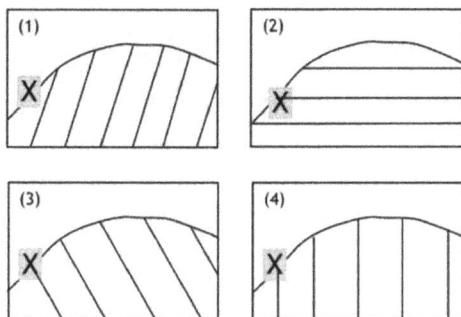

If a road is to be cut into the hill at point X, which of
these bedding types would be the least stable and so
unsafe for road construction?

   A. 1
   B. 2
   C. 3
   D. 4

9. The following diagram shows an electrolysis cell for
the refining of copper:

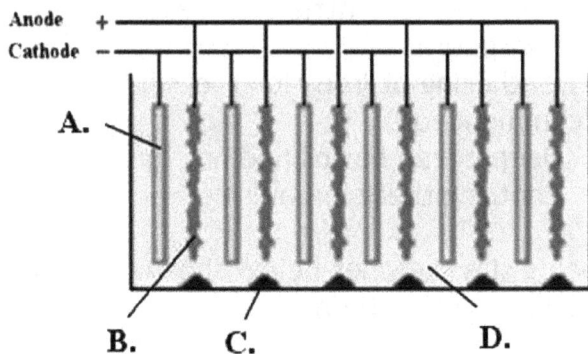

Diagram showing the Electro-refining of Copper

The place where pure copper is deposited is:

A. A
B. B
C. C
D. D

10. Of the currently available energy technologies, the one which may have potential to have a long-term supply of fuel with minimal emission of greenhouse gases is:

A. Coal steam systems
B. Nuclear breeder reactors
C. Oil-fired steam systems
D. Nuclear fusion reactors

Review and Discussion Questions

1. Briefly distinguish between covalent and ionic bonds. Give an example of each in minerals.

2. What is chemically important about the element silicon? Why does it form so many compounds?

3. Name a mineral which has:

    (a) hardness 7 and conchoidal fracture
    (b) hardness 3 and gives carbon dioxide gas
        when acid is dropped onto its surface
    (c) dark-coloured mineral with hardness of
        5 to 6 and two cleavages at about 60°

4. Given two minerals in fine powder form and which each have a gold colour and metallic lustre, how could one quickly determine whether the sample is gold or pyrite (fool's gold)?

5. The minerals calcite (calcium carbonate) and barite (barium sulfate) can look identical (having the same crystal shape, cleavage, colour, hardness). How could one distinguish between them?

6. Explain why quartz and diamond are very hard minerals.

7. A good specimen of a mineral can be scratched with quartz but it will also scratch apatite. What is the most likely hardness of this new mineral? How could you confirm this conclusion?

8. Discuss the importance of ionic substitution in producing a wide range of minerals

9. Research the Internet about the use of minerals:

        (a) in the home
        (b) in the office
        (c) as gemstones and
        (d) in international trade.

10. Discuss the factors which could make a mineral one of economic importance. Could a mineral change its status of being economically important or not?

11. Define each of the following terms:

        (a) goaf
        (b) gangue
        (c) overburden
        (d) fissionable

12. Briefly explain, using appropriate mineral/ore examples how each of the following is formed:

        (a) hydrothermal ore bodies
        (b) alluvial deposits
        (c) orthomagmatic deposits
        (d) gossans.

13. A large gold deposit has been located at a depth of 400 metres within jointed metamorphic rock approximately 5 km from a large town. The population of the town is approximately 4000 persons and the surrounding country is mainly used for cattle grazing. The country consists of rolling plains with the occasional rocky outcrop and a small river runs through the town. About one quarter of the country is covered in light forest with an abundance of native wildlife.

The mining company is discussing whether or not to use open cut, deep shaft or a chemical mining method such as borehole mining using the cyanide technique. Critically evaluate the potential hazards of the suggested methods prior to giving your decision by:

    (a) identifying the factors which would be potentially hazardous due to mining

    (b) evaluating which mining technique would be most appropriate for the mine by giving at least two reasons for your choice and

    (c) critically evaluating the hazards and other concerns to the environment and the community (give at least three items and how they could be overcome)

14. Countries such as Australia have huge deposits of uranium minerals which are mined and exported to other countries, yet Australia does not have a nuclear power industry. Use the Internet to review the advantages and disadvantages of nuclear power generation and its possible usefulness in the future.

15. Do a power audit of your own home over one week (use the meter box to estimate the amount of kilowatt hours used). Are there ways in which you can save energy?

     (a) at home
     (b) using transportation
     (c) at work
     (d) in your neighbourhood

What are some alternatives which could be used in these places? What could be done in these places if the total mains (grid) electricity system is shut down?

Answers to Multichoice Questions

Q1. B Q2. B Q3. B Q4. A Q5. C
Q6. C Q7. D Q8. A Q9. A Q10. B

Reading List

Anthony, J. W., Bideaux, R. A., Bladh, K. W. & Nichols, Monte C.,(Edits).(1995). *Handbook of Mineralogy.* Chantilly, VA, US: Mineralogical Society of America. ISBN: 978-0-9622097-1-0

Barthelmy, D. (2015). *Mineralogy Database.* (an extremely useful website) http://webmineral.com/

Bullock, L. & Hustrulid, William A.(Edits.). (2001). *Underground Mining Methods: Engineering Fundamentals and International Case Studies.* Littleton, Colo. Society for Mining, Metallurgy, and Exploration. 718 pp. ISBN: 0873351932.

Cox, D. P. & Singer, D. A. (*Editors*). 2014? *Mineral Deposit Models.* USGS Publication. http://pubs.usgs.gov/bul/b1693/html/bull1nzi.htm.

Davenport, W. G., King, M., Schlesinger, M. and Biswas, A.K. (2002). *Extractive Metallurgy of Copper, 4th Edition.* Oxford, England, Pergamon Press pp 454. ISBN:0080440290.

Deer, W.A., Howie, R.A. & Zussman, J. (1992). *An Introduction to the Rock Forming Minerals* (2nd edition ed.). London: Longman ISBN: 0-582-30094-0

Dill, H.G. (2010). *The "Chessboard" Classification Scheme of Mineral Deposits: Mineralogy and Geology from aluminum to zirconium.* Earth-Science Reviews Volume 100, pp. 1-420.

Dixon, C. J. (2012). *Atlas of Economic Mineral Deposits.* New York.Springer Science & Business Media, pp 139. ISBN: 9789401165112

Dyar, M.D. & Gunter, M.E. (2008). *Mineralogy and Optical Mineralogy.* Chantilly, Virginia: Mineralogical Society of America. ISBN: 978-0-939950-81-2.Gem.

Gemological Institute of America. (2002). *Diamonds and Diamond Grading: The Evolution of Diamond Cutting.* Gemological Institute of America, Carlsbad, California. https://www.gia.edu/

Geology.com. (2015). Mineral Uses, Properties & Descriptions. (another great database) http://geology.com/minerals/

Gribble, C.D. & Read, H.H. (1989). *Rutley's Elements of Mineralogy,* 27th Edition. London: Allen & Unwin, pp 560. ISBN: 9780045490110

Guilbert, John M. and Park, Jr. Charles F. (1986). *The Geology of Ore Deposits,* New York. W. H. Freeman, pp 985. ISBN: 0-7167-1456-6

Haldar, S.K. (2013). *An Introduction to Mineralogy and Petrology.* Atlanta, GA.: Elsevier Science Publishing, pp 356. ISBN: 9780124167100

Harlow, P.G.L. (1994). *Gems*. Ipswich, QLD: Pedrick, pp 33. ISBN: 0 646 13241 5

Hartman, H. L. & Mutmansky, J. M. (2002). *Introductory Mining Engineering*. Hobocan, NJ. John Wiley & Sons, pp 570.

Hefferan, K. & O'Brien, J. (2010). *Earth Materials*. NY: Wiley-Blackwell, pp 608.

International Energy Agency (IEA): http://www.iea.org/topics/electricity/

Kraus, P. D. (1987). *Introduction to Lapidary. Iola, Wisconsin: Krause Publications. p. ix*. ISBN 978-0-8019-7266-9.

Klein, C. & Philpotts, A. (2012). *Earth Materials – Introduction to Mineralogy and Petrology*. Cambridge University Press. 552 pages. ISBN: 9780521145213

Kundu, K. & Kumar, A. (2014). Biochemical Engineering Parameters for Hydrometallurgical Processes: Steps towards a Deeper Understanding. *Journal of Mining: Volume 2014 (2014), Article ID 290275, 10 pages*

Mayer, W. (1991). *A Field Guide to Australian Rocks, Minerals and Gemstones*. Sydney: Ure Smith, pp 335. ISBN:0 7254 0816 2.

National Environment Agency (NEA): http://www.nea.gov.sg/

Nelson, S.A. (2014). *Earth Structure, Materials, Systems, and Cycles.* Tulane University, Dept. of Earth & Environmental Sciences: EENS 3050 http://www.tulane.edu/~sanelson/Natural_Disasters/struct&materials.htm

Pohl, Walter L. (2011). *Economic Geology: Principles and Practice.* New York. John Wiley & Sons, pp 680. ISBN: 978-1-4443-3663-4

Renewable Policy Energy Network in the Twenty-first Century (REN 21): http://www.ren21.net/wpcontent/uploads/2016/06/GSR_2016_Full_Report_REN21.pdf

adit (93) is a horizontal tunnel going into a mine from outside. if sloped or spiralled it is called a decline and the entrance is called a portal.

allotrope (38) is a different physical forms of the same element due to differences in crystal structure e.g. carbon as graphite or diamond.

alloys (39) are mixtures of two or more metals e.g. electrum, a mixture of gold, silver and copper.

alluvial (67) deposits consist of weathered material containing valuable minerals deposited downhill by water.

alpha particles α (185) consist of a helium atomic nucleus of two neutrons and two protons. These have total electrical charge of +2 and an atomic mass of 4 units compared to the mass of one proton = 1.

aluminium smelting (147) is the complex process of converting bauxite to pure aluminium metal.

amorphous (13) having no crystals at all.

amphiboles (46) are inosilicates having a double chain of silicate tetrahedra.

amygdales (111) are spherical gas bubble holes which may be partly or completely filled with fine crystals of quartz or more solid chalcedony.

anaerobic (151) is any process which happens without air, such as conversion of plant matter to coal.

anions (34) are negatively-charged ions having gained electrons.

anode copper (138) is the almost pure copper collected on the negative terminal in an electro-refining cell.

Archimedes' Principle (29) states that the upward buoyant force that is exerted on a body immersed in a fluid, whether fully or partially submerged, is equal to the weight of the fluid that the body displaces and it acts in the upward direction at the centre of mass of the displaced fluid.

asterism (105) in gemstones is a star-like pattern of radiating light known as asterism when view in good light from above.

atomic mass (185) is the mass of the nucleus of an atom and is due to the number of protons and neutrons in its centre (nucleus). For a comparison, the basic mass is that of the hydrogen atom which is taken as 1.

atomic number (180) is the number of protons in the centre or nucleus, of an atom which also equals the number of electrons in its surrounding shells.

atomic pile (184) is an old term for the older cores of nuclear fission plants using piles of graphite blocks.

atmosphere(1) consists of the gases of the air, mostly nitrogen gas (78% volume), oxygen gas (21%), argon gas (less than 1%), carbon dioxide gas (0.03%), water (varies) and small traces of other gases, dust and pollen;

autoclaves (183) are high temperature and pressure vessels used in processing and sterilizing.

Banded Iron Formations BIFS (64) are sedimentary deposits of finely layered alternations of silica and iron minerals such as haematite (iron oxide Fe_2O_3), magnetite (iron oxide Fe_3O_4), or siderite (iron carbonate $FeCO_3$).

basal cleavage (20) is perfect cleavage in one direction.

Basic Oxygen Steelmaking BOS (145) uses liquid pig-iron from the blast furnace and scrap steel, and accounts for over 60% of the world's steel production.

batter (75) is the inclined wall of a bench or horizontal step in an open pit mine. The edge of the bench where the angle changes from steep to horizontal is called a berm.

baveno twinning (14) is a form of simple twinning, named after the town in northern Italy in which the crystals are joined together in a diagonal pattern e.g. orthoclase feldspars

benches (75) are horizontal slopes found naturally or recessed into an open pit or hillside in an open-cut mine to be used as a road or to stabilize a deposit.

berm (75) is a sudden change or flattening of the angle of a beach or mined slope.

beta particles β⁻ (185) are high-speed, negatively-charged electrons and have very low mass.

biofuels (218) are fuels which have been derived from biomass or biological material such as agricultural products, wastes from sugar and timber mills, gases coming from landfill dumps and other waste gases.

biomass (218) is any agriculture or plant matter used in making biofuels.

blister copper (138) is the intermediate form of copper produced from smelting and has a blister-like surface when cooled.

boghead coal (156) also called torbanite, is a black coal with a high algal content formed in lacustrine conditions. it is named after a small village in Scotland.

bord (100) is the space or room left when coal is mined underground leaving pillars to support the roof.

Borehole Mining BHM (90) involves the drilling of boreholes into a subsurface body and extraction of the wanted resource using high pressure water and sometimes air to bring the material to the surface.

breeder reactors (191) are nuclear reactors which produce more fissionable material than they consume to generate energy.

brunted (122) is the act of rounding of diamonds by agitating one diamond with another.

BTEX (168) is a mixture of benzene, toluene, ethylbenzene and xylene with sodium hypochlorite, hydrochloric acid, cellulose, acetic acid and disinfectants use in borehole mining of coal seam gas.

cabachon (105) is a round-topped gemstone with a flat under surface.

cannel coal (156) is a non-banded, hard coal with a small algal content and probably formed in smaller pools. Its name is a corruption of candle as it burns with a bright, soot-free flame.

carat (101) is a weight = 0.2 grams and named from the ancient weight of a carob seed used in trading.

cast iron (144) is impure iron produced by refining crude or pig iron from the blast furnace. It contains carbon and is therefore brittle. It is often used to cast iron products from molds.

cations (34) are ions having a positive charge, having lost electrons.

caving (94) is an underground mining technique where a fractured ore body is mined from below allowing the ore to fall down into the working area.

centrifuge (183) is a high speed spinning device used in separating materials of different weights.

chain reaction (184) is when the neutrons released by fission collide with other nuclei in other atoms and cause more fission, more neutrons, radioactivity and large amounts of heat.

chemistry (29) is the composition of a mineral in terms of its constituent atoms or ions and their rations.

christmas tree (164) in the oil industry, this is a multiple-tube valve used to connect to the end of an oil well to take the crude oil to different places.

clastic sedimentary rocks (4) have been formed by the settling or compaction of minerals or organic matter and show no visible sediments.

clasts (4) are the particles of sediment e.g. sand, gravel etc. found in sedimentary rocks.

clay (8) is a fine-grained mineral or complex derived mainly from the weathering of feldspars and similar alumina-silicates.

cleavage (19) is the way that some minerals split along natural planes (flat surfaces) of weakness when gently struck.

coalification (151) is the process which turns compressed vegetation into coals with the loss of water (H_2O), carbon dioxide (CO_2) and methane (CH_4).

coke (140) is refined carbon used in industry such as in iron production and is made by the distillation of coal i.e. by heating coal in the absence of air.

colour (15) is the wavelength of light emerging from within the internal structure of the mineral – other wavelengths of the visible spectrum being absorbed.

colluvial (67) material containing valuable minerals which is carried downhill as simple dry sediment and deposited.

composite stone (120) is a gemstone prepared as several layers of the original stone cemented onto other material to add bulk and enhance the stone. it may have two layers as a doublet or three as a triplet.

Concentrated Solar Power CSP (207) is an industrial application of solar heating which generates a very large amount of solar by using mirrors or lenses to concentrate a large area of sunlight onto a small area.

conchoidal (22) fracture in minerals is seen as concentric circles at a sharp edge and is named after the fracture seen in broken shells.

contact metamorphic rocks (5) are formed from existing rocks which have been changed by the heat of a nearby igneous body.

copper matte (137) is the mixture of metal sulfides which fall to the floor of the furnace in the smelting phase of metal production.

copper smelting (136) is the process which converts copper ores to an impure form of copper using a furnace, oxygen and fluxes.

country rock (93) is a general term for surrounding rock which may contain an igneous intrusion, ore body or other structure.

covalent bonding (37) consists of atoms held together by sharing outer electrons.

critical mass (184) is the amount of a fissionable material needed to start a chain reaction.

crystal family (23) is a mineral classification based upon three-dimensional axes imagined through a perfect crystal of the mineral.

crystal lattice (35) is the framework of ions which make up the shape of the crystal.

crystal system (27) this is a more technical classification of crystals and involves planes, or mirrors of symmetry and numbers of rotation around each axis as a rotational symmetry.

crystallization (2) is the solidification of materials into regular forms.

cubic (25) crystal family in which the axes are equal length and are all at 90° to each other e.g. halite (common salt).

cyclone (127) is a separating device consisting of a spril shute down which water and rock and ore are passes with the heavier ore being separated by centifigal force.

cyclosilicates (45) from the greek cyclos meaning a ring, these consist of multiples of rings formed by six tetrahedra linked together by sharing two common oxygen atoms.

daughter products (181) are the several isotopes formed when an atom is split by fission.

decline (93) is a long sloping tunnel into an underground mine.

deuterium (187) is an isotope of hydrogen containing one proton, as in normal hydrogen, plus a neutron within the nucleus.

diaphaneity (17) is the way that light passes through the specimen in normal thickness or not.

diapir (163) is a structureformed when salt pushes up underground forming a dome.

dredging (87) is carried out from floating vessels either on natural waterways over the ore or on man-made ponds over mineral-rich sands.

Electric Arc Furnace EAF (146) is used in steelmaking using scrap steel or iron already reduced by natural gas as the main feed materials. High voltage electrical energy is then used to melt the solid iron with air to remove any carbon.

electrolysis (149) is the splitting up of water (or other solutions) by electricity.

elluvial (67) consists of weathered material containing valuable minerals which is deposited at the original site of weathering or nearby to it.

evaporite deposits (66) are formed through the evaporation of mineral-rich solutions.

extrusive igneous rocks (4) are formed on the earth's surface from crystallization from lava.

faceting (121) is the act of cutting stones with several flat faces at appropriate angle to the type of cut of the stone.

fermentation (218) is the natural process of converting sugars into ethanol alcohol using yeasts.

flux (145) an additive added to smelting processes to aid melting.

fracking (168) or hydraulic fracturing is the breaking up of rock to extract coal seam gas using solutions which are pumped down into the rock.

fracture (22) is the way the mineral breaks up but not along smooth planes of weakness – conchoidal (shell-like) being the most useful.

froth flotation (127) is the separation of two wanted ore powders by the use of agitated froth of water and chemicals which selectively remove one ore from the top froth.

fuel cells (175) are a type of electrical battery which produce electricity when certain fuels (e.g. oxygen and hydrogen) are added as fuels.

gamma radiation γ (185) consists of high energy electromagnetic radiation similar to x-rays but more penetrating having no mass but great penetrating ability.

gangue(61) refers to any waste minerals which is not wanted in a process.

gem (100) is any material, natural or man-made used for adornment.

gemstones (100) are minerals and rocks which may be used for adornment.

geochemistry (33) is a major sub-division of geology which involves the testing and determination of the chemical composition of minerals in a laboratory.

geodes (111) or "thunder eggs" are spheres of quartz material, often hollowed in their middle formed by in-filling of gas bubbles with silica solution.

geosequestration (175) is the removal of carbon dioxide gas from a system by pumping it deep below ground into a salt-water (saline) aquifer.

geothermal energy (219) is energy produced using the heat of the earth.

goafs (100) are at the end of an underground coal mine where the roof is allowed to cave in under control for further excavation of the coal.

gossans (59) are raised, circular structure of hardened, oxidised mineral found on the surface as a cap on the ore body below.

habit (11) describes the overall appearance and arrangement of the crystals of the mineral specimen.

half-life (181) is the time for a radioactive material to decay to half of its original radioactivity.

hardness (18) is the resistance to scratching when tested with a standard set (Mohs' Scale) of items having a relative strength. common quartz has a hardness of 7.

heavy water (187) is used as the moderator to slow down the neutrons of the chain reaction and consists of water molecules (H_2O) in which the hydrogen part is a natural isotope called deuterium, which has an extra neutron in its nucleus.

heft (29) is the relative heaviness of a mineral.

hexagonal (25) crystal family with two horizontal axes equal in length and at 120^0 to each other and an unequal vertical axis at 90^0 to these two e.g. beryl (beryllium aluminium silicate).

Hilt's Law (152) the rank of coal increases with depth if the thermal gradient or increase in temperature with depth, is entirely vertical.

hydraulicking (87) utilizes a high-pressure stream of water which is directed against the mineral deposit to remove it by the erosive action of the water.

hydrosphere (1) is the earth's layer of water which covers over 70% of the planet's surface as ocean, ice caps, fresh water and soil moisture; and the humidity.

hydrothermal deposits (60) are formed from hot circulating water-rich fluids containing dissolved minerals.

igneous rocks (3) are formed from crystallization from molten rock material either from magma or lava.

inosilicates (45) from the Greek *inos* for fibre, have single or double chains of silicate tetrahedra.

intrusive igneous rocks (4) are formed from molten magma below the surface.

ionic bonds (35) are formed from the joining of charged atoms or groups by electrostatic attraction.

ionic substitution (40) occurs when ions of similar size will replace others within the crystal lattice.

ions (34) are atoms which have gained an electrostatic charge by losing or gaining negatively-charged electrons from their outer shell.

iron smelting (140) is the conversion of iron ores such as haematie to impure iron within a blast furnace using coke, limestone and oxygen.

isotope (180) is a different form of the same chemical element due to different amounts of neutrons (neutrally-charged particles) within their nucleus (atomic centres) i.e. they have the same atomic number but different atomic mass.

kelly (164) is an oil drilling term for the hexagonal or square pipe attached to a rotating table driven by a motor on an oil rig.

kimberlite (60,102) is a type of carbon-rich igneous rock found in the pipes of ancient volcanoes and often found as a source of diamonds.

lacustrine environments (151) are those of still water lakes.

lapidary (123) is the art of cutting, polishing and setting gemstones.

lava (2) is molten rock flowing upon the surface.

lignocellulosic (219) refers to the dry matter of plants, mostly as the long-chain molecules of cellulose and lignin.

lithosphere(1) consists of rocks, minerals , soils and weathered products which make up the more solid material of the outer parts of the earth's crust.

longwall (159) mining is used in coal mines to mine along a face of the coal seam using an excavator which runs along on a rail system.

lustre (16) is the way which light reflects off the surfaces of the mineral specimen. Lustre may be metallic or non-metallic.

magma (2) is molten rock below the surface of the earth.

metallic bonds (39) are those found in metals which easily lose 1,2 or 3 electrons from their outer shell.

metalliferous ores (57) are minerals of value mined containing metals within their composition and mined so that the metals can be extracted, e.g. galena (lead sulfide).

metamorphic rocks (5) are formed from existing rocks by changes due to heat and/or pressure.

minerals (2) are the basic components of the lithosphere and are naturally-occurring, inorganic (i.e. non-living) substances having definite properties and a known chemical composition.

Mississippi Valley Type MVT (63) deposits are found at the base of limestone within a marine sedimentary sequence such as that found in the Mississippi River valley, where hydrothermal solutions had penetrated the original strata, cooled and reacted with the limestone forming galena (lead sulfide) and sphalerite (zinc sulfide).

moderator (187) is used to slow down neutrons in a reactor and thus control the chain reaction.

monoclinic (26) crystal family in which all axes are unequal in length with one axis vertical, another at 90^0 to the vertical and the third axis at an oblique angle to the plane of the other two axes e.g. muscovite.

multiple twinning (14) occurs when many crystals grow side-by-side many times with a reversal of crystal growth direction e.g. plagioclase feldspar.

nesosilicates (44) from the Greek *nesos* meaning island, have lattices made up of separated (islands) silicate tetrahedra linked by positive metal ions.

neutrons (180) are sub-atomic particles within the nucleus of atoms and has mass but no electrical charge.

non-clastic sedimentary rocks (4) do not have visible grains or crystals.

nuclear fission (180) is the splitting of an atom's nucleus by high-speed neutrons with energy and particle release.

nuclear fusion (191) is the joining together at very high temperatures of lighter atomic nuclei to form heavier nuclei with the emission of energy and radioactive particles.

nucleus (180) is the centre of an atom.

octohedral cleavage (20) is cleavage perfectly in four directions giving pyramid shapes.

open pit (75) open cut or open cast mining, is a surface mining technique used to exploit a near-surface deposit.

ore (3,56) are minerals of economic value being a source of metal or other resource.

ore genesis (59) of ore bodies is the processes within the earth which forms bodies of ore.

Organic Rankine Cycle ORC (222) is a dual (binary) system which uses low temperature heat sources to heat, vaporise and expand an organic fluid which then drives a turbine couples to an electric generator.

orthomagmatic deposits (59) ores formed from magmas and found within large igneous intrusions.

orthorhombic (26) is a crystal family in which all axes are unequal in length but are all at 90^0 to each other e.g. aragonite.

overburden (75) is unwanted rock removed from above economic deposits mined by open pit and strip mining.

paludal environments (151) are those of stagnant swamps.

panning (89) is an old technique used in small scale individual separating of valuable materials such as gold and relies on the relative heaviness of the wanted material compared to the soil or gravel in which it is contained.

pegmatitic deposits (60) are formed from magmas containing dissolved water and so crystallised slowly forming large crystals include some which are valuable e.g. feldspar and mica.

penetration twinning (14) in which one crystal grows through the other at an angle e.g. staurolite.

phosphate deposits (69) are those rocks rich in the element phosphorus e.g. phosphorite.

photoelectric effect (208) occurs when certain high frequency wavelengths of light (i.e. their photons) strike a metal or semiconductor and knock off electrons to the surrounding vacuum.

photons (208) are extremely small amounts or quanta, of light energy which act like solid particles.

Photovoltaic Effect PV (208) is when high frequency wavelengths (or photons) strike a semi-conductor and produce a voltage (electrical pressure) to allow electrons to flow within the material as an electrical current.

phyllosilicates (46) from the Greek *phylon* meaning leaf, have tetrahedra linked by three shared oxygen atoms giving extensive two-dimensional sheets.

pig iron (140) is the impure form of crude iron produced from a blast furnace. It is so called because the ingots or slabs of cooled iron produced reminded the workers of pigs seen from the side.

placer deposits (67) are formed by gravity taking weathered material containing the valuable minerals downhill either by water (alluvial) or as simple dry sediment slides (colluvial). some material may also remain at the site of weathering or nearby to it (elluvial).

placer mining (87) is the mining of loose alluvium containing valuable minerals by removing the sediment using hydraulicking (powerful jets of water), dredging (scooping up with conveyed buckets) and panning (agitation within water flowing over a grooved surface) and separating it from the ore.

plasma (192) consists of very hot, electrically charged gas such as that found in the sun and in electrical discharges.

pleochroism (117) in a mineral is when it will change colour at different angles of light when the cut stone is rotated.

portals (93) are the entrances to a decline or long tunnel into an underground mine.

precious gemstones (101) are relatively rare and usually have very high value, usually over $5000 per carat.

Pumped Hydro Energy Storage PHES (202) is a system whereby excess electricity at off-peak times is used to pump water back uphill into a storage dam so that it can be released later at peak times to produce needed electricity.

pyroxenes (45) are inosilicates which have a single chain of silicate tetrahedra.

quanta (208) are small bundles of energy considered to act as particles.

ranks (151) are the grades of coal ranging from compressed vegetation (peat) to metamorphosed carbon (anthracite).

rare earths (58 Table 3.1) are those chemical elements with atomic numbers ranging from 57 to 71 and which are of value and have similar properties. they are often found together in ores and include: lanthanum, cerium, praseodymium, neodymium, promethium, samarium, europium, gadolinium, terbium, dysprosium, holmium, erbium, thulium, ytterbium, lutetium, scandium and yttrium. they have a wide range of uses from magnets to lenses and electronics.

regional metamorphic rocks (5) are formed from existing rocks over a large area due to extreme pressure and heat.

regolith (10) is a general name for broken and loose material such as sands, soils, gravel, etc.

resonance (39) is the oscillation or switching of bonds from single to double within ring molecule or crystal groups.

rhombohedral (28) is a crystal system often used synonymously for the trigonal system, but is defined by its crystal lattice rather than outward appearance and is a subset of the trigonal lattice system.

rocks (3) are two or more minerals in combination by crystallization or cemented together.

sand and gravel (8) are larger clasts which have weathered out of the rock. These clasts are often rounded. Gravel may also consist of rounded rock fragments.

scree (10) is the name given to loose, broken material deposited by gravity down the sides of cliffs.

sedimentary rocks (4) are formed from loose, eroded material which has been compacted and cemented.

Sedimentary Exhalative Deposit SEDEX (62) deposits are formed by release of the hot ore-bearing fluids into the ocean along with the deposition of the usual marine sediments.

semi-precious gemstones (108) are the more common gemstones, usually (but not always) of lower value.

silicates (42) are an extensive family of minerals consisting of the elements silicon and oxygen and combined with other elements. Silica (as quartz) is the simplest member as SIO_2.

simple twinning (13) is when two crystals grow together in some symmetrical way e.g. side-by-side or as a mirror image e.g. gypsum

slag (137) the waste material manufactured within various furnaces so that it can be removed from the process of smelting.

smelting (9) is the extraction of a metal from its ore, usually involving chemical reactions and considerable heat.

soil (10) is a mixture of minerals, organic matter, gases, liquids, and countless organisms that together support life on Earth.

solar electricity (208) is the production of domestic and industrial electricity using the rays of the sun.

Solar Thermal Energy STE (205) uses solar collectors for heating water.

sorosilicates(44) from the Greek *soros* meaning a group, these consist of groups of two linked tetrahedra with metal ions within the lattice.

specific gravity (28) is the ratio of the density of a mineral (its mass divided by its volume) to the density of water (which =1.0 g/cc).

specific properties (30) of minerals are unique features of that mineral which sets it apart from others.

steel (144) is processed iron which contains less carbon than cast iron. It often has other metals added to it to form special steel for construction and industry.

sticks (164) are the hollow pipes used in oil drilling. they are able to be screwed together and inserted into the hexagonal fitting (kelly) within the rotating table which drives the drill stick. the drill bit with cutting teeth is attached to the end of the stick.

stope (94) is a vertical area of collapse formed by blasting the ore body. It is often then mined from below.

stratiform ore bodies (63) consist of layers of ores, usually sulfides of copper, lead and zinc sulfides within rock strata.

streak (15) is the colour of the powdered mineral when scratched.

strip mining (86) uses a shallow, open cut mine in which sections of the surface are mined in strips with previously excavated sections being refilled with the waste material mined.

tailings (83) consist of fine powdered mined ore usually mixed with water as a slurry and piped into a tailings dam where the water is evaporated.

tectosilicates (47) from the Greek *tecto* meaning framework, these silicates form complex, three-dimensional frameworks in which all oxygen atoms at the apices or points of the tetrahedra are shared.

tetragonal (25) crystal family in which two axes are equal in length and the other shorter or longer, with all axes at 90 degrees e.g. zircon.

tidal energy (212) is a form of hydropower which uses the movements of the earth's tides to produce electricity.

Tidal Stream Generators TSG (214) make use of fast flow moving water in streams to directly power turbines, in a similar way to wind turbines and using similar rotor types.

triclinic (27) crystal family in which all axes are unequal and none of the angles are at 90^0 e.g. plagioclase.

trigonal (27) is a crystal system with all axes equal and all angles equal (but not 90^0). It is a subset of the hexagonal system e.g. quartz.

tsunami (195) is an ocean wave generated by an earthquake with a large vertical component.

twinning (13) in minerals occurs when crystals grow together.

unconformity (163) where there has been a break in sedimentation and where strata may meet at an angle with impervious rocks overlying oil reservoir rocks.

van der Waal's bonds (38) are weak forces between molecules or sheets of minerals due to the attraction of the electrons of one set of atoms to the nuclei of another.

valency (42) or valency is the combining ability of atoms.

Volcanogenic Massive Sulfide deposits VMS (62) deposits are formed on the ocean floor by circulating hydrothermal fluids from volcanic vents called black smokers.

wave action generators (215) use the power of ocean waves created by winds and currents to generate electricity.

winze (93) is a short vertical connection (often by ladder) between two levels in an underground mine.

yellowcake (182) is the refined powdered uranium oxide as (U_3O_8) concentrate.

This book is also available in electronic format which can be purchased at amazon.com for Kindle or other electronic devices such as PCs and iPad using the free Kindle App. Printed books in the series **ADVENTURES IN EARTH SCIENCE** are available from Felix Publishing, Australia (info@felixpublishing.com) and include:

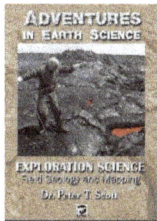

EXPLORATION SCIENCE
Field Geology & Mapping

FOSSILS - LIFE in the ROCKS

RICHES from the EARTH
Minerals & Energy

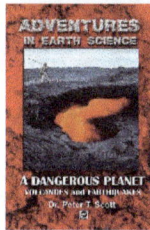

A DANGEROUS PLANET
Volcanoes & Earthquakes

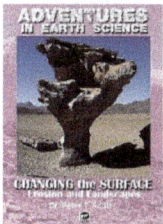

CHANGING the SURFACE
Erosion & Landscapes

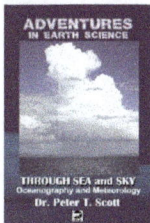

THROUGH SEA and SKY
Oceanography & Meteorology

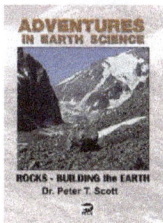

ROCKS - BUILDING the EARTH

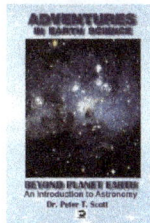

BEYOND PLANET EARTH
An Introduction to Astronomy